MATH WORD PROBLEMS

David Ebner, Ph.D.

THE EASY WAY

All inquiries should be addressed to:
Barron's Educational Series, Inc.
250 Wireless Boulevard
Hauppauge, New York 11788
http://www.barronseduc.com

ISBN-13: 978-0-7641-1871-5
ISBN-10: 0-7641-1871-4

Library of Congress Catalog Card No. 2001043558

Library of Congress Cataloging-in-Publication Data
Ebner, David, 1942–
    Math word problems the easy way / David Ebner.
        p.   cm.—(Barron's easy way series)
    Includes index.
    ISBN 0-7641-1871-4
        1. Mathematics—Problems, exercises, etc.   2. Word problems
(Mathematics)   I. Title.   II. Series.
QA43 .E26 2002
510′.76—dc21   2001043558

Printed in the United States of America
9 8 7 6 5 4

Contents

Introduction

World Problems the Easy Way includes exercises ranging from algebra through calculus. Each example is clearly dissected and analyzed and all the material is presented concisely.

All the examples are broken down into the following areas:

1. Initial problem,

2. Analysis of the problem,

3. Work area,

4. Answer.

The analysis area is probably the most important section because we discuss the general approach to take to solve the problem. The work area shows the "nuts and bolts" operations.

The philosophy of this book is to present the student with a mathematics "cookbook" with step-by-step solutions for each example.

This book is unique in that each exercise is linked to an earlier example.

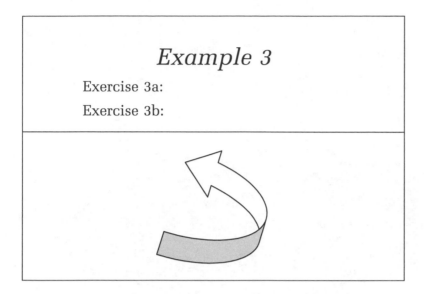

Example 3

Exercise 3a:

Exercise 3b:

The book thus comes with its own built-in tutor.

Each chapter includes a test. The solutions to the exercises and the tests are found in the appendix, as is a table of values of trigonometric functions and a glossary.

1
Formulas

A **formula** is a convenient shorthand means of expressing relationships among variables. In most cases, a formula makes life much simpler for us. The formula allows us to predict outcomes. It is not an explanation of the phenomenon.

Once a formula has been developed, we can mechanically substitute values. There are, literally, thousands of formulas used every day in business and science.

1.1 What Is a Formula?

Example 1

Jamal has just purchased a new car and wants to determine its fuel efficiency—the number of miles per gallon the car can travel. To do this, he must divide the number of miles driven by the number of gallons used by the engine. Help Jamal develop a formula to determine the car's fuel efficiency.

Analysis

Let mpg = the number of miles per gallon.

Let m = the number of miles driven.

Let g = the number of gallons of gasoline used.

To obtain the number of miles per gallon, divide the number of miles, m, by the number of gallons, g.

Work

$$\text{mpg} = \frac{m}{g}$$

Answer

$$\text{mpg} = m/g$$

Exercises

1. In each case, develop a general formula.

1a. The volume of a rectangular solid, V, equals length, l, times width, w, times height, h.

1b. The perimeter of a rectangle, P, equals two times the length, l, plus two times the width, w.

1c. The mean, M, of four numbers, a, b, c, and d, equals their sum divided by 4.

1d. Jackie's salary, s, is determined by multiplying the number of hours worked, h, by his hourly wage, w.

Example 2

In Example 1, if Jamal drives the car 4,190.4 miles and the car uses 194 gallons of gasoline, how many miles per gallon does the car get?

Analysis

Let m = the mileage driven.

Let g = the number of gallons of gasoline used.

Let mpg = the number of miles per gallon.

Just substitute the given figures directly into the formula and then solve for mpg.

Work

$$\text{mpg} = \frac{m}{g}$$

$m = 4{,}190.4$, $g = 194$: $\text{mpg} = \dfrac{4{,}190.4}{194} = 21.6$

Answer

mpg = 21.6 miles per gallon

---

2. Use the figures given here and substitute into the formulas developed for the corresponding problems in Exercise 1.

2a. Find the volume, V, of a rectangular solid whose length = 4.5 inches, width = 8 inches, and height = 7.2 inches.

2b. Determine the perimeter of a rectangle, P, whose length is 9 feet and whose width is 4.6 feet.

2c. Find the mean, M, of 78, 95, 66, and 81.

2d. Determine Jackie's salary, s, if he works 38 hours at $15.48 per hour.

1.3 Changing the Subject of a Given Formula

Let's say that we have two flasks, A and B; flask A is one ounce heavier than flask B. Is there another way we can present this information? Sure! We can say that flask B weighs one ounce less than flask A.

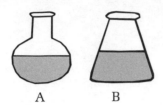

A   B

$$A = B + 1 \quad \text{or} \quad B = A - 1$$

In the inimitable words of Albert Einstein, "Everything is relative!"

Similarly, we can change the subject of any mathematical formula.

Example 1

The total cost of repairing a car, C, is obtained by adding the cost of the parts, k, plus $60 times the number of hours worked, h. First develop a formula directly from the given information. Then rearrange this formula in order to find k in terms of C and h.

Analysis

Let C = the total cost.

Let k = the cost of the parts.

Let h = the number of hours worked.

Start with the given formula and then rearrange it so that the cost of the parts, k, is alone on one side.

Work

$$C = k + 60h$$

Subtract $60h$: $C - 60h = k$

Symmetric Property: $k = C - 60h$

Answer

$$k = C - 60h$$

Exercises

1a.   The cost of taking a taxi ride, c, is determined by multiplying the number of miles driven, m, by $2.20 added to $3. Develop a formula for this relationship. Then, find m in terms of c.

1b.   The perimeter of a triangle, P, is found by adding the three sides, a, b, and c. Express the relationship in a formula and then find a in terms of P, b, and c.

1c.   The amount of money accumulated in a bank account, A, is determined by adding the initial principal, p, to the interest, i. Develop a formula to express the relationship among all the variables and then find i in terms of the other variables.

Example 2

The volume, V, of a right circular cone equals $(\pi r^2 h)/3$. Express the value of h in terms of r and V.

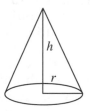

Analysis

Let V = the volume of a cone.

Let r = the radius of the cone.

Let h = the height of the cone.

Develop a formula for the volume of a right circular cone and then rearrange the formula so that h is alone on one side of the equation.

Work

$$V = \frac{\pi r^2 h}{3}$$

Multiply by 3:

$$3\left(V = \frac{\pi r^2 h}{3}\right)$$

$$3V = \pi r^2 h$$

Divide by πr^2:

$$\frac{3V}{\pi r^2} = \frac{\pi r^2 h}{\pi r^2}$$

$$\frac{3V}{\pi r^2} = h$$

Symmetric Property:

$$h = \frac{3V}{\pi r^2}$$

Answer

$$h = \frac{3V}{\pi r^2}$$

2a.   The circumference of a circle, C, is found by multiplying 2π by the radius, r. Develop a formula for finding the radius, r, in terms of C.

2b.   The volume of a right circular cylinder, V, is found by taking the product of π, the radius squared, r^2, and the height, h. Find h in terms of the other variables.

1.4 The Celsius/Fahrenheit Relationship

In the United States and in English-speaking countries, temperature is generally measured on the Fahrenheit scale. The rest of the world measures heat in Celsius (or Centigrade). The Swedish astronomer Anders Celsius developed the Celsius scale. It actually makes more sense than the Fahrenheit scale because there are 100 degrees between the freezing point of water (measured at standard atmospheric pressure) at 0° Celsius and its boiling point at 100° Celsius. The German physicist Gabriel Fahrenheit devised the Fahrenheit scale in which the freezing point of water (also at standard atmospheric pressure) is 32° Fahrenheit while its boiling point is 212° Fahrenheit.

We frequently want to convert from one scale to another, so we need a convenient formula to help us with the conversion. The formula $C = (5/9)(F - 32)$ shows us the relationship between Celsius and Fahrenheit and allows us to change from one scale to the other.

Example 1

Change 77° Fahrenheit to Celsius.

Analysis

Let F = the Fahrenheit temperature.

Let C = the Celsius temperature.

We'll use our formula and then simply substitute 77 for F.

Work

$$C = (5/9)(F - 32)$$
$$F = 77: \quad C = (5/9)(77 - 32)$$
$$C = (5/9)(45)$$
$$C = 25$$

Answer

25° Celsius

At times we may want to convert from Celsius to Fahrenheit, so it's convenient to be able to convert in either direction.

1a.   Yuan is conducting a chemistry experiment, but his manual describing the procedure indicates the Fahrenheit temperature as 131°, and he only has a Celsius thermometer. Help Yuan change 131° Fahrenheit to Celsius.

1b.   The meteorologist predicts that the average temperature for today is going to be 104° Fahrenheit. Change that to Celsius.

Exercises

Example 2

Find the Fahrenheit equivalent of 55° Celsius.

Analysis

Let F = the Fahrenheit temperature.

Let C = the Celsius temperature.

Use the same formula and then just substitute 55 for Celsius.

Work

$$C = (5/9)(F - 32)$$

$C = 55$:
$$55 = \frac{5}{9}(F - 32)$$

Multiply by $\frac{9}{5}$:
$$\frac{9}{5}\left(55 = \frac{5}{9}(F - 32)\right)$$

$$99 = F - 32$$

Add 32:
$$131 = F$$

Answer

131° Fahrenheit

2a.   Millie is on vacation is Spain and doesn't feel well. She takes her temperature and jumps when she sees that the thermometer reads 35° Celsius. Millie lives in the United States and is used to reading Fahrenheit temperatures. She really doesn't know whether she has a fever, so she has to use the Celsius/Fahrenheit formula to determine her temperature in Fahrenheit. Find Millie's temperature on the Fahrenheit scale.

Exercises

2b.   A certain microchip in a computer cannot be subject to temperatures below 10° Celsius. What is this in Fahrenheit?

1.5 Simple Interest

Simple interest is interest earned on the principal invested. Compound interest, on the other hand, is interest earned on principal plus interest earned on past interest. It's much easier to determine simple interest than compound interest.

To find simple interest, I, we just multiply principal, P, by rate per year, R, by the time period (in terms of a year), T.

$$I = PRT$$

The total amount of money remaining in a bank account, A, after a number of simple interest additions equals the initial principal, P, plus the simple interest, I:

$$A = P + I$$

Since we know that $I = PRT$, if we need to, we can substitute PRT for I so that this formula can also be written as

$$A = P + PRT$$

Example 1

Find the simple interest, I, earned on a principal of $1,400 at an annual interest rate of $6\frac{1}{2}\%$ for 3 years.

Analysis

Let I = the interest.

Let P = the principal.

Let R = the rate per year.

Let T = the number of years.

$6\frac{1}{2}\% = 0.065$. Use the formula $I = PRT$ and then substitute values for the variables.

Work

$$I = PRT$$

$P = 1,400, R = 0.065, T = 3$: $I = (1,400)(0.065)(3)$

$$I = 273$$

Answer

$273

1a.   Rosa wants to take out a loan of $3,600 from the Gotham National Bank. If the bank charges a simple annual rate of interest of 9%, how much interest will she have to pay at the end of a three-year loan?

1b.   Ng has an account at the Federal Bank of New Hampshire. He bought a Certificate of Deposit that paid a simple interest rate. If he invested $4,000 for a period of five years at a simple annual interest rate of 5%, how much interest was earned at the end of the time period?

Example 2

Josie places $2,400 in a bank account for a period of two years and six months at a simple annual interest rate of 4.7%. How much is in the account at the end of the time period?

Analysis

Let A = amount of money in the account after two years and six months.

Let P = the principal.

Let R = the rate per year.

Let T = the number of years.

4.7% = 0.047; 2 years, 6 months = 2.5.

We want to find the *total amount* in the bank account, A, so use the formula $A = P + PRT$.

Work

$$A = P + PRT$$

$P = 2,400,$
$R = 0.047, T = 2.5:$ $A = 2,400 + (2,400)(0.047)(2.5)$

$$A = 2,400 + 282$$
$$A = 2,682$$

Answer

$2,682

2a.   Morrie placed $2,300 in a bank paying a simple annual interest rate of $4\frac{1}{2}$%. If he kept the money in the bank for four years, how much did he have at the end of that time?

2b. Makeba purchased a $4,000 U.S. Treasury Note that paid a simple annual interest rate of 5.3%. If she kept the note for $4\frac{1}{2}$ years, how much money did she have?

Example 3

Max opened a bank account with an initial deposit of $1,900. If he leaves the money in the bank for four years and nine months and, at the end of this period, has $2,351.25 in his account, what simple rate of interest did the bank pay?

Analysis

Let A = the amount of money in the account after four years and nine months.

Let P = the principal.

Let R = the rate per year.

Let T = the number of years.

Start off with the formula for the total amount of money in the account, $A = P + PRT$. Then substitute values for A, P, and T to find the rate of interest, R.

Work

$$A = P + PRT$$

$A = 2{,}351.25$, $P = 1{,}900$,

$$T = 4\frac{9}{12} = 4\frac{3}{4} = 4.75:$$

$$2{,}351.25 = 1{,}900 + (1{,}900)(R)(4.75)$$

$$2{,}351.25 = 1{,}900 + 9{,}025R$$

Subtract 1,900: $451.25 = 9{,}025R$

Divide by 9,025:

$$0.05 = R$$

Answer

5%

3a. Sharon just saw her bank statement, which indicated that she had $5,832 in her account. If she initially deposited $5,400 for two years, what was the simple annual rate of interest?

3b. If the original principal in a bank account was $600, the amount accumulated over a number of years was $690, and the annual rate of interest was 5%, for how many years was the principal invested?

To find the distance traveled, d, just multiply the rate, r, by the time in motion, t. Let's now apply the formula to some practical situations.

Example 1

Raquel leaves River City and drives to Gotham at an average speed of 62 mph. If the trip takes her five hours and 15 minutes, how far apart are the cities?

Analysis

Let d = the distance driven.

Let r = the rate of speed.

Let t = the time for the trip.

5 hours 15 minutes = $5\frac{15}{60} = 5\frac{1}{4} = 5.25\,\text{hr}$

Use the distance formula $d = rt$.

Work

$$d = rt$$

$r = 62, t = 5.25$: $\quad d = (62)(5.25) = 325.5$

Answer

325.5 miles

1a. Jacqueline drove at an average speed of 61 mph for 6.3 hours. How far did she drive?

1b. Mike left college and boarded a bus for home. If the bus driver drove at an average rate of 64.6 mph for seven hours until Mike arrived home, how long was the trip?

Example 2

A Blue Line Bus leaves Parkersville and travels to Narrowsburg at an average speed of 59.5 mph. If the distance between the cities is 357 miles, how long should the trip take?

Analysis

Let d = the distance driven.

Let r = the rate of speed.

Let t = the time for the trip.

Use the distance formula $d = rt$; then substitute values for d and r.

Work

$$d = rt$$

$d = 357$, $r = 59.5$:	$357 = 59.5t$
Divide by 59.5:	$6 = t$
Symmetric Property:	$t = 6$

Answer

6 hours

Exercises

2a. The distance between two points is 420 miles. If Crystal drives at an average rate of 56 mph, how many hours should the trip take?

2b. Julio is the train engineer. If the distance between two cities is 350.4 miles and he wants to make the trip in six hours, what should his average rate of speed be?

Example 3

Tony leaves his home at 10 A.M. and drives to his sister's house. If he drives at an average speed of 53 mph and the distance between their homes is 238.5 miles, at what time will he arrive?

Analysis

Let d = the distance driven.

Let r = the rate of speed.

Let t = the time for the trip.

Find the number of hours it took Tony to drive the distance and then add the answer to 10 A.M.

Work

$$d = rt$$

$d = 238.5, r = 53$: $238.5 = 53t$

Divide by 53: $4.5 = t$

Symmetric Property: $t = 4.5$

Add 4.5 hours to
10 A.M.:   10 A.M. + 4.5 hours = 2:30 P.M.

Answer

2:30 P.M.

3a.   Mr. Saunders is a salesperson. He has to leave Silver City at 11:30 A.M. and travel 250 miles to Pico. If he drives at an average speed of 62.5 mph, at what time should he arrive in Pico?

3b.   Kendra leaves her home at 1 P.M. If she drives at an average speed of 54 mph and her destination is 405 miles away, at what time should she arrive?

Let's see whether we can determine a pattern in the following design.

First, we'll assign a number corresponding to the number of symbols in each of the rows: 2, 5, 8, 11, 14. Is there any formula we can develop to show a pattern to these numbers? Yes, each number is three more than the immediately preceding number.

Row # in Diagram, n	# of Symbols	Relationship Between Row # and # of Symbols
1	2	$2 = 2 + 0(3) = 2 + (n - 1)3$
2	5	$5 = 2 + 1(3) = 2 + (n - 1)3$
3	8	$8 = 2 + 2(3) = 2 + (n - 1)3$
4	11	$11 = 2 + 3(3) = 2 + (n - 1)3$
5	14	$14 = 2 + 4(3) = 2 + (n - 1)3$

In this table, let n = the number of the nth term in the sequence (Row #), a_1 = the first term, a_n = the nth term, and d = the common difference between terms. Since all the differences between succeeding terms are the same, we can select any two succeeding terms and obtain the difference.

To develop a formula for any term—call it the nth term, a_n—we'll multiply $(n - 1)$ times the difference between any two terms, d, and add that product to the first term, a_1:

$$a_n = a_1 + (n-1)d$$

Let's use this formula to check the numbers in the right-hand column in this table:

$$a_n = a_1 + (n - 1)d$$

First term: $a_1 = 2 + (1 - 1)(3) = 2 + 0(3) = 2$

Second term: $a_2 = 2 + (2 - 1)(3) = 2 + 1(3) = 5$

Third term: $a_3 = 2 + (3 - 1)(3) = 2 + 2(3) = 8$

Fourth term: $a_4 = 2 + (4 - 1)(3) = 2 + 3(3) = 11$

Fifth term: $a_5 = 2 + (5 - 1)(3) = 2 + 4(3) = 14$

Sequence

A **sequence** is an ordered set with a first element and succeeding elements. The terms of the sequence are the individual members of the set:

$$a_1, a_2, a_3, \ldots$$

An **Arithmetic Progression (A.P.)** is a special kind of sequence derived by *adding a constant number* to each term to arrive at a successor term. In our example, the Arithmetic Progression is 2, 5, 8, 11, 14. The constant 3 is added to each term to arrive at the next term.

Example 1

Charisse is a stockbroker. Her company wants her to increase sales each week by $500. If she sold $4,000 in stocks the first week, how much should she sell in the 12th week?

Analysis

Let a_1 = the first term in the Arithmetic Progression.

Let n = the number of terms.

Let a_n = the nth term in the A.P.

Let d = the difference between terms.

We have an A.P. and we want to find the amount of sales in the 12th week, so we'll use the formula $a_n = a_1 + (n - 1)d$ to find her required sales in the 12th week.

Work

$$a_n = a_1 + (n - 1)d$$

$a_1 = 4{,}000$, $n = 12$,
$d = 500$:

$$a_{12} = 4{,}000 + (12 - 1)\,500$$

$$a_{12} = 4{,}000 + 11(500)$$
$$= 4{,}000 + 5{,}500 = 9{,}500$$

Answer

$9,500

1a.   Hilda lifts weights. She wants to increase the number of pounds she lifts each week by 6 pounds. If she starts lifting 74 pounds, how much should she be able to lift in the eighth week?

1b.   Kelly is a sales representative. His supervisor wants him to increase sales by $400 per week. If Kelly now sells $12,600 per week, how much should he sell in the 14th week from now?

Example 2

Hailey bought a new car and is slowly breaking it in by increasing her mileage by a constant number of miles each week. How many miles will she have to drive the car in the 11th week if she drove it 280 miles in the third week and 400 miles in the sixth week?

This problem is a little more difficult than Example 1, so we need to break it up into two parts.

Analysis, Part I

First, we must find the difference between terms, d. To do this, we'll drop the first two terms from the original A.P. and develop a new A.P.

Original A.P.:   _, _, 280, _, _, 400, _, _, _, _, _

New A.P.:   280, _, _, 400, _, _, _, _, _

In the new A.P., 280 is the first term and 400 is the fourth term.

Let a_1 = the first term in the Arithmetic Progression.

Let n = the number of terms.

Let a_n = the nth term in the A.P.

Let d = the difference between terms.

Work

Let's find the difference, d:

$$a_n = a_1 + (n - 1)d$$
$$a_4 = a_1 + (n - 1)d$$

In the new A.P., $a_1 = 280$, $a_4 = 400$, $n = 4$:

$$400 = 280 + (4 - 1)d$$
$$400 = 280 + 3d$$

Subtract 280: $120 = 3d$

Divide by 3: $40 = d$

Symmetric Property: $d = 40$

The difference between terms is 40.

Analysis, Part II

Now let's find the ninth term in the new A.P. (the 11th term in the old A.P.).

Work

$$a_n = a_1 + (n - 1)d$$

In the new A.P., $a_1 = 280$, $n = 9$, $d = 40$:

$$a_9 = 280 + (9 - 1)(40)$$
$$a_9 = 280 + 8(40)$$
$$a_9 = 600$$

a_9 in the new A.P. is the same as a_{11} in the old A.P.

Answer

600 miles

Exercises

2a. The Harris School has a fundraising campaign. The school wants to increase its collections by a constant number of dollars per week. The school raises $563 during the third week and $605 during the sixth week. If everything goes according to schedule, how much should be raised during the 12th week?

2b. Little Mike is training for the state fair hot dog–eating contest. He wants to increase the number of hot dogs he eats every day by a constant number. If he eats six hot dogs on the second day and 12 hot dogs on the fourth day, how many hot dogs should he eat on the seventh day?

A series is a sequence with addition signs placed between terms:

$$a_1 + a_2 + a_3 + \cdots$$

An **Arithmetic Series** is an Arithmetic Progression with addition signs inserted between terms. For example, in example 2, the arithmetic series is

200 + 240 + 280 + 320 + 360 + 400 + 440 + 480 + 520 + 560 + 600

To find the sum of an arithmetic progression, we get the average of the first and last terms and then multiply by the number of terms.

$$S = \frac{n(a+l)}{2}$$

where S = sum of the arithmetic series, a = first term, l = last term, n = number of terms.

Example 3

Let's go back to stockbroker Charisse and find the total of her sales for the first 12 weeks.

Analysis

Let S = the sum of the terms in an A.P.

Let n = the number of terms.

Let a = the first term in the A.P.

Let l = the last term in the A.P.

We want the total of her sales for all 12 weeks, so we'll use the formula for the total of an A.P.

$$S = \frac{n(a+l)}{2}$$

Work

She sold $4,000 in stock the first week and $9,500 the 12th week, so $a = 4{,}000$, $n = 12$, and $l = a_{12} = 9{,}500$.

$$S = \frac{12}{2}(4,000 + 9,500) = 6(13,500) = 81,000$$

Answer

$81,000

Exercises

3a. Krishna is saving his money. The first week he saves $16. Each succeeding week he saves $2 more than the preceding week. What is the total amount he has saved after 30 weeks?

3b. The Humongous Computer Corporation wants to increase its sales by an additional $50,000 each week. If it sells $80,000 worth of products the first week, what is the target for total sales for 15 weeks?

1.8 Geometric Progression

Let's try to determine a pattern in the following design.

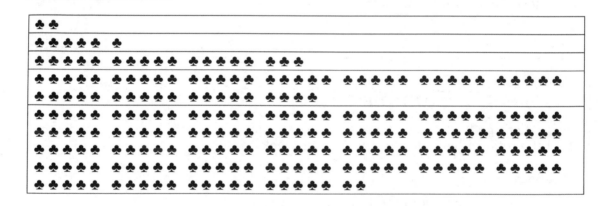

First, we'll assign a number corresponding to the number of symbols in each of the rows: 2, 6, 18, 54, 162. Is there any formula we can develop to show a pattern to these numbers? Yes, each number is three times the immediately preceding number.

Row # in Diagram, n	# of Symbols	Relationship Between Row # and # of Symbols
1	2	2
2	6	$6 = 2(3) = 2(3)^1 = 2(3)^{2-1}$
3	18	$18 = 2(9) = 2(3)^2 = 2(3)^{3-1}$
4	54	$54 = 2(27) = 2(3)^3 = 2(3)^{4-1}$
5	162	$162 = 2(81) = 2(3)^4 = 2(3)^{5-1}$

In this table, let n = the number of the nth term in the sequence (Row #), a_1 = the first term, a_n = the nth term, and r = the common ratio.

2, 6, 18, 54, 162 is called a Geometric Progression.

The terms of a G.P. may be represented as a_1, a_2, a_3, . . . , a_n.

To find the common ratio, r, divide any term by its immediate predecessor:

$$r = \frac{a_n}{a_{n-1}}$$

To find the common ratio, we can take any term, say the fourth term, a_4, and divide it by its predecessor, a_3. So, in this example,

$$\frac{a_4}{a_3} = \frac{54}{18} = 3$$

The common ratio is, therefore, 3. This means that we must multiply the preceding term by 3 to get to the next term in the progression.

Using this table, let's see whether we can develop a formula to find any specific term in a Geometric Progression:

First term: $\quad a_1 = \quad 2 = 2(3)^0 = 2(1) = a_1$

Second term: $\quad a_2 = \quad 6 = 2 \times 3 = 2(3)^1 = a_1 r^1$

Third term: $\quad a_3 = \quad 18 = 2 \times 3 \times 3 = 2(3)^2 = a_1 r^2$

Fourth term: $\quad a_4 = \quad 54 = 2 \times 3 \times 3 \times 3 = 2(3)^3 = a_1 r^3$

Fifth term: $\quad a_5 = 162 = 2 \times 3 \times 3 \times 3 \times 3 = 2(3)^4 = a_1 r^4$

If we take the first term, a_1, and multiply it by r^{n-1}, where r = the common ratio and n = the order of the term, we can derive any term in the progression:

$$a_n = a_1 r^{n-1}$$

A **Geometric Progression (G.P.)** is a special kind of sequence. A G.P. is derived by multiplying each term by a constant number—called the **common ratio**—to arrive at a successor term.

Example 1

A certain strain of bacteria is multiplying fourfold every day. If the bacteria weigh 0.003 grams to begin, how much will the bacteria weigh in 8 days?

Analysis

Let a_n = the nth term in the Geometric Progression.

Let a_1 = the first term in the G.P.

Let r = the common ratio.

Let n = the number of terms in the G.P.

This is a perfect example of a G.P., so let's find the weight of the bacteria on the eighth day, a_8.

Work

$$a_n = a_1 r^{n-1}$$

$a_1 = 0.003$, $r = 4$,
$n = 8$:

$$a_8 = 0.003(4)^{8-1} = 0.003(4)^7$$
$$= 0.003(16{,}384) = 49.152$$

Answer

49.152 grams

Exercises

1a.   Jack has an investment. If his money increases threefold every year and he starts with $4, how much will he have after six years?

1b.   Virus X237 doubles every four hours. If there are five viruses to start, how many viruses are there after 32 hours?

Example 2

Cameron City has a projected population of 186,624 in 10 years from now. If its population increases geometrically at the rate of 20% every two years, what is its population right now?

Analysis

Let a_n = the nth term in the Geometric Progression.

Let a_1 = the first term in the G.P.

Let r = the common ratio.

Let n = the number of terms in the G.P.

Since this is a G.P., we'll use the G.P. formula to find the first term, a_1, of the progression. In this case, we'll first find the common ratio and then work backwards in order to determine the initial population.

If the population increases at a rate of 20% every two years, then we can multiply our initial population by 120% to determine the population two years from now. The initial population is a_1.

In two years the population will be 20% more than the first year's population, so that makes the second term in the Geometric Progression 120% (a_1).

We can clearly see that the common ratio is 120%, or 1.20. Now we want to find the current population, a_1, so we'll use the formula $a_n = a_1 r^{n-1}$. Since the population increases by 20% every two years, in 10 years there will be five terms in the G.P., so $n = 5$.

Work

$$a_n = a_1 r^{n-1}$$

$r = 1.20, n = 5$: $\quad a_n = a_1 r^{5-1}$

$a_5 = 186,624$: $\quad 186,624 = a_1(1.20)^4$

$$186,624 = a_1(2.0736)$$

$$\frac{186,624}{2.0736} = a_1$$

$$90,000 = a_1$$

Answer

90,000

2a. The Xcel Corporation has increased its sales by 30% each year. If the corporation now has annual sales of $25,990,510, what were its sales six years ago?

2b. Upon descending, a minisubmarine takes on water and increases its weight at the rate of 10% every five minutes. If the submarine now weighs 1,317,690 pounds, how much did it weigh at the moment of descent 25 minutes ago?

A **Geometric Series** is a Geometric Progression with addition signs inserted between terms.

Let's determine the formula for the sum, S, of the geometric series

$$a_1 + a_1 r^1 + a_1 r^2 + a_1 r^3 + \cdots + a_1 r^{n-1}$$

Let S = the sum:

(1) $\quad S = a_1 + a_1 r^1 + a_1 r^2 + a_1 r^3 + \cdots + a_1 r^{n-1}$

Multiply by r:

(2) $\quad rS = \quad\quad a_1 r^1 + a_1 r^2 + a_1 r^3 + \cdots + a_1 r^{n-1} + a_1 r^n$

Subtract: (2) from (1):

$$S - Sr = a_1 \quad\quad\quad\quad\quad\quad - ar^n$$

Factor:

$$S(1 - r) = a_1 \quad\quad\quad\quad\quad\quad - ar^n$$

Divide by $(1 - r)$:

$$S = \frac{a_1 - a_1 r^n}{1 - r}$$

Or, if we multiply numerator and denominator by -1:

$$S = \frac{(-1)(a_1 - a_1 r^n)}{(-1)(1 - r)} = \frac{-a_1 + a_1 r^n}{-1 + r} = \frac{a_1 r^n - a_1}{r - 1}$$

Example 3

Let's solve the following English nursery rhyme:

As I was going to St. Ives,
I met a man with seven wives;
Every wife had seven sacks,
Every sack had seven cats,
Every cat had seven kits.
A man, kits, cats, sacks, and wives,
How many people did I meet?

Analysis

Let S = the sum of the terms in a Geometric Progression.

Let a_1 = the first term in the G.P.

Let r = the common ratio.

Let n = the number of terms in the G.P.

The rhyme is a Geometric Progression, so we need to use the formula for the sum of a G.P. The first term is 7, the ratio is 7, and there are four terms—wives, sacks, cats, and kits.

Work

$$S = \frac{a_1 r^n - a_1}{r - 1}$$

$a_1 = 7$, $r = 7$, $n = 4$:

$$S = \frac{7(7)^4 - 7}{7 - 1} = \frac{16{,}807 - 7}{6} = 2{,}800$$

Answer

$$2{,}800 + 1 \text{ man} = 2{,}801$$

3a.   Jack picks up bottles and returns them for the deposit money. On his first day, he picks up 13 bottles. If he triples his collection each day, what is the total number of bottles he picks up in five days?

3b.   Hanchen is on the track team. She plans on doubling her running distance every week. During her first week, she runs 2 miles. How many miles does she run altogether in the six weeks?

1.9 Scientific Formulas

   Most of the formulas we have thus far reviewed have been "consumer oriented." However, in today's technological world, it is increasingly critical for all of us to have some general understanding of scientific formulas. In the next two examples, let's just try to understand the general concepts.

Example 1

   The tip speed of an airplane propeller, TS, as it rotates around the engine equals πdR. Find the tip speed when the diameter of the propeller is 70 inches and it makes 800 revolutions per second.

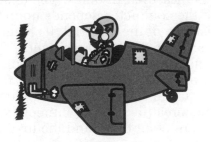

Analysis

   Let TS = tip speed of the propeller.

   Let $\pi = 3.14$.

   Let d = the diameter of the propeller.

   Let R = the revolutions per second the propeller makes.

Use the given formula and just substitute the values for π, d, and R.

Work

$$TS = \pi dR$$

$\pi = 3.14$, $d = 70$, $R = 800$: $TS = (3.14)(70)(800)$

$$TS = 175{,}840$$

Answer

The tip speed of the propeller is 175,840 inches per second.

Exercises

1a.   Boyle's Law, developed by English scientist Robert Boyle, states that the pressure of a gas times its volume is equal to a constant number, for a gas at a constant temperature. In mathematical terms, Boyle's Law is represented as $pv = k$, where $p =$ pressure, $v =$ volume, and $k =$ the constant. Find k when $p = 67\,\text{lb/in.}^2$ and $v = 460\,\text{in.}^3$.

1b.   The total resistance, R_T, in an electrical circuit with two resistances, R_1 and R_2, equals $(R_1 \times R_2)/(R_1 + R_2)$. Find the total resistance when $R_1 = 16\,\text{ohms}$, $R_2 = 12\,\text{ohms}$. Round your answer to the nearest tenth of an ohm.

Example 2

The resistance of an electrical conductor is its obstruction to the flow of electricity, and this obstruction is measured in ohms. The total resistance of the conductor depends upon the specific resistance of the material, its length, and its cross-sectional area. The total resistance, R, of the conductor equals the product of the specific resistance of the metal (in ohms), k, and the length of the conductor (in centimeters), L, divided by the cross-sectional area (in square centimeters), A. To the nearest tenth of an ohm, find the total resistance of a conductor when the specific resistance is 0.49, the length is 140 centimeters, and the cross-sectional area is 32 square centimeters.

Analysis

Let $R =$ the total resistance of the conductor.

Let $k =$ the specific resistance of the conductor.

Let $L =$ the length of the conductor.

Let A = the cross-sectional area of the conductor.

Develop a formula and then substitute values for k, L, and A.

Work

$$R = \frac{kL}{A}$$

$k = 0.49$, $L = 140$,

$A = 32$:
$$R = \frac{(0.49)(140)}{32} = 2.14375 \approx 2.1$$

Answer

2.1 ohms

The total resistance (in ohms) of an electrical conductor, R, equals kL/A, where k = the specific resistance of the material (in ohms), L = the length of the material (in centimeters), and A = the cross-sectional area (in square centimeters).

2a. Find R when $k = 0.38$, $L = 115$ centimeters, and $A = 28$ square centimeters. Round your answer to the nearest hundredth.

2b. Find R when $k = 0.59$, $L = 320$ centimeters, and $A = 41$ square centimeters. Round your answer to the nearest hundredth.

Exercises

Example 3

Find the cross-sectional area of a conductor when the total resistance is 7.8 ohms, the specific resistance is 0.06 ohms, and the length of the conductor is 52 centimeters.

Analysis

Let R = the total resistance of the conductor.

Let k = the specific resistance of the conductor.

Let L = the length of the conductor.

Let A = the cross-sectional area of the conductor.

Use the given formula and just substitute values for R, k, and L.

Work

$$R = \frac{kL}{A}$$

$R = 7.8$, $k = 0.06$, $L = 52$:
$$7.8 = \frac{(0.06)(52)}{A}$$

Multiply by A:
$$A\left(7.8 = \frac{(0.06)(52)}{A}\right)$$

$$7.8A = 3.12$$

Divide by 7.8:
$$A = 0.4$$

Answer

0.4 square centimeters

Exercises

3a.   Find A when $R = 19.008$ ohms, $k = 36$, and $L = 132$ centimeters.

3b.   Find L when $A = 12$ square centimeters, $k = 45$, and $R = 112.5$ ohms.

Test

1.   The profit at a school dance, P, is represented by the formula $P = \$12.50T - \$75C$, where $T =$ the number of tickets sold and $C =$ the number of chaperones. What is the profit when 300 tickets were sold and there were eight chaperones?

2.   Shaquan's take-home pay, T, is given by the formula $T = 0.75(\$13.50h)$, where $h =$ the number of hours worked. Find his take-home pay if he works 38 hours.

3.   The average of three numbers, A, is found by adding all the numbers, a, b, and c, and then dividing by 3. Express c in terms of a, b, and A.

4.   The volume of a right circular cylinder, V, is found by taking the product of π, r^2, and h, where $r =$ the radius and $h =$ height. Express h in terms of π, r^2, and h.

5–6.   Use the formula $C = (5/9)(F - 32)$, where $C =$ Celsius and $F =$ Fahrenheit.

5.   Change 77° Fahrenheit to Celsius.

6.   Find the Fahrenheit equivalent to 110° Celsius.

7.   Find the simple interest earned on a principal of $5,600 invested at a simple annual interest rate of $7\frac{1}{2}\%$ for 4 years. Use the formula $I = PRT$, where $I =$ interest,

P = principal, R = annual rate of interest, and T = time, in years.

8–9. Use the formula $A = P + PRT$, where A = the amount in the bank account at the end of the time period, P = principal, R = annual rate of interest, and T = time, in years.

8. How much money is in a bank account if an initial principal of $8,900 is left for five years and nine months at a simple annual interest rate of 4%?

9. At the end of three years and six months, Joe has $5,687 in his bank account. If he opened the account with $4,700, how much simple annual interest did the bank pay?

10. Enriqués drives from Central City to Leander at an average rate of speed of 56 mph. If the trip takes six hours and 30 minutes, what is the distance between the two cities?

11. If a 747 airliner flew at an average speed of 583 mph for seven hours and 15 minutes, how many miles did it fly?

12. Makeba leaves her home at 9:00 A.M. and drives to her vacation place. If she drives at an average rate of 61 mph for a distance of 335.5 miles, at what time will she arrive at her destination?

13–14. To find the nth term in an arithmetic progression, use the formula $a_n = a_1 + (n - 1)d$, where n = the number of terms, a_n = the nth term, a_1 = the first term, and d = the difference between terms.

13. Hanna is a deep-sea diver. She wants to increase her time underwater by five minutes on each successive dive. If she stayed underwater for 32 minutes on her first dive, how long should she stay underwater during the 11th dive?

14. Joseph wants to increase his savings by a constant amount each week. If he saves $16 the second week and $43 the fifth week, how much will he have to save during the 13th week?

15. Julio is on the track team. He has increased his running distance by a constant amount each time on the track. He started off by running 1,000 meters and increased this amount by 300 meters each time he ran. If he ran 12 times, what was his *total* running distance? Use the formula $S = (n/2)(a + l)$, where S = the sum, n = the number of occurrences, a = the first term, and l = the last term in an arithmetic progression.

16–17. Use the formula $a_n = a_1 r^{n-1}$, where a_n = the nth term in the Geometric Progression, a_1 = the first term, and r = the common ratio.

16. Virus 345A multiplies threefold each day. If the viruses weigh 0.002 grams on the first day, what is their total weight on the seventh day?

17. The number of cases of skin cancer is increasing geometrically at the rate of 30% a year. In four years, the number of cases of skin cancer in one section of the United States is projected to be 2,570,490. How many cases of skin cancer are there currently in the area?

18. Mr. Bradley has a savings plan at work. Each year he is going to save three times the amount he saved the previous year. If, at the end of 8 years, he has saved a total of $39,360, how much did he save the first year? Use the formula $S = (a_1 r^n - a_1)/(r - 1)$, where S = sum, a_1 = the first term in the Geometric Progression, r = the common ratio, and n = the number of terms in the G.P.

19. The resistance of an electrical conductor is given by the formula $R = (kL)/A$, where R = the total resistance of the conductor (in ohms), k = the specific resistance of the conductor (in ohms), L = the length of the conductor (in centimeters), and A = the cross-sectional area of the conductor (in square centimeters). Find R when $k = 0.59$, $L = 96$, and $A = 48$.

20. The tip speed of an airplane propeller, TS, as it rotates around the engine is given by the formula πdR, where d = the diameter of the propeller (in inches) and R = revolutions per second. Find the diameter of the propeller if $TS = 141,300$ inches per second, $\pi = 3.14$, and $R = 900$.

2

Mixture and Coin Problems

Remember those movies with a mad scientist mixing her chemicals in a lab for all kinds of nefarious purposes? If she didn't mix those chemicals properly, she would blow herself up rather than the rest of the world, so she had to be super-correct in her mixes.

Besides scientists, numerous other people, including bakers, candy-makers, bankers, and economists should know how to combine various ingredients in order to obtain a desired mixture. Local bakers and candy-makers, of course, don't actually sit down and figure out the best mix for their product. However, you can be sure that large bakeries and candy companies pay close attention to the proper mix for their products, for a small slip in the mix may affect the taste of the candy or cake or may end up costing the corporation thousands of dollars.

Similarly, bankers and economists must mix their own type of product—bonds, certificates of deposit, etc.—to put together financial packages subject to certain conditions.

2.1 Mixing Foods

Candy manufacturers often offer mixes of candies to their buyers. The manufacturer must factor in the original prices of the separate candies to determine the price of the mixture.

Linear equations are used to solve mixture problems. In general, the best way to construct this type of linear equation is to set the total price of the mix equal to the totals of the individual candies.

Example 1

The Big C Candy Corporation sells chocolate kisses at $0.22 per ounce. How much is the cost of 15 pounds?

Analysis

One pound equals 16 ounces, so let's determine the number of ounces in 15 pounds. After we find out how many ounces we're talking about, we'll just multiply the number of ounces by $0.22 per ounce.

Work

$$15 \text{ pounds} = 16 \times 15 \text{ ounces} = 240 \text{ ounces}$$

$$240 \text{ ounces} \times \$0.22 \text{ per ounce} = \$52.80$$

Answer

$52.80

Exercises

1a.   Cashew nuts sell for $0.42 an ounce. What is the cost of 12 pounds?

1b.   The Big Peach Supermarket is selling raisins for $0.20 an ounce. How much do 4.5 pounds cost?

Example 2

A grocer mixes 10 pounds of Brazilian coffee at $4.20 per pound with per 20 pounds of Venezuelan coffee at $7.20 per pound. How much is one pound of the new mixture?

Analysis

The total weight of the mixture is $10 + 20 = 30$ pounds.

Let x = the price per pound of the new mixture.

Remember, number of pounds multiplied by price per pound equals the total price.

Kind	Number of Pounds	Price Per Pound ($)	Total Price = Pounds × Price ($)
Brazilian	10	4.20	(10)(4.20)
Venezuelan	20	7.20	(20)(7.20)
Mixture	30	x	$30x$

Work

The total price of the Brazilian coffee added to the total price of the Venezuelan coffee equals the price of the mixture.

$$10(4.20) + 20(7.20) = 30x$$
$$42.00 + 144.00 = 30x$$

Multiply by 10:
$$10(42 + 144 = 30x)$$
$$420 + 1,440 = 300x$$
$$1,860 = 300x$$

Divide by 300:
$$6.20 = x$$

Symmetric Property:
$$x = 6.20$$

Check

$$10(4.20) + 20(7.20) = 30x$$

$x = 6.20$:
$$42 + 144 = 30(6.20)$$
$$186 = 186 ✓$$

Answer

$6.20 per pound

2a. Jaroslav mixes 16 pounds of chocolates at $3.60 a pound with 24 pounds of mints at $4.20 a pound. How much should he sell a pound of the mixture for?

2b. Twenty-one pounds of scrap metal at $1.40 a pound are mixed with 28 pounds of scrap metal at $3.50 per pound. Find the price of 1 pound of the new mixture.

Example 3

Ms. Chisholm mixes chocolate mints selling for $8.96 a pound with jelly rings selling for $4.48 a pound to obtain a mixture selling for $7.28 a pound. There are 8 pounds more of chocolate mints than there are of jelly rings. How many pounds of each kind of candy are there?

Analysis

Let x = the number of pounds of jelly rings.

Let $x + 8$ = the number of pounds of chocolate mints.

Kind	Number of Pounds	Price Per Pound ($)	Total Price = Pounds × Price ($)
Jelly rings	x	4.48	$4.48x$
Chocolate mints	$x + 8$	8.96	$8.96(x + 8)$
Mixture	$2x + 8$	7.28	$7.28(2x + 8)$

Work

The total price of the chocolate mints plus the total price of the jelly rings equals the total price of the mixture:

$$4.48x + 8.96(x + 8) = 7.28(2x + 8)$$

Multiply by 100:

$$100(4.48x + 8.96(x + 8) = 7.28(2x + 8)$$

$$448x + 896(x + 8) = 728(2x + 8)$$

$$448x + 896x + 7,168 = 1,456x + 5,824$$

$$1,344x + 7,168 = 1,456x + 5,824$$

Subtract 1,456x:

$$-112x + 7,168 = 5,824$$

Subtract 7,168: $-112x = -1,344$

Divide by 128: $x = 12\,(\text{jelly rings})$

$x + 8 = 20\,(\text{chocolate mints})$

Check

$$4.48x + 8.96(x+8) = 7.28(2x+8)$$

$x = 12:$ $4.48(12) + 8.96(20) = 7.28(32)$

$$53.76 + 179.20 = 232.96$$

$$232.96 = 232.96\,\checkmark$$

Answer

12 pounds of jelly rings and 20 pounds of chocolate mints

3a. To obtain a mixture at $2.40 a gallon, $2.48 a gallon gasoline is mixed with $2.24 a gallon gasoline. If there are five fewer gallons of the less expensive gasoline than of the more expensive type, determine the number of gallons of each kind of gasoline.

3b. Cashews at $7.40 a pound are mixed with peanuts at $2.60 a pound. If there are 18 pounds in the mixture and the mixture sells for $5.80 per pound, how many pounds of cashews are there in the mixture?

Example 1

An adult ticket to a movie is $7.50, while a child's ticket is $2.70. If a total of 280 people were admitted to a theater and the receipts amounted to $1,677.60, how many tickets of each type were sold?

Analysis

Let x = the number of adult tickets.

Let $280 - x$ = the number of children's tickets.

Kind	Number of Tickets	Price Per Ticket ($)	Total Price = Number × Price ($)
Adult	x	7.50	$7.50x$
Child	$280 - x$	2.70	$2.70(280 - x)$

Work

The total price of adult tickets plus the total price of children's tickets equals $1,677.60:

$$7.50x + 2.70(280 - x) = 1,677.60$$

Multiply by 100:

$$100(7.50x + 2.70(280 - x) = 1,677.60)$$
$$750x + 270(280 - x) = 167,760$$
$$750x + 75,600 - 270x = 167,760$$
$$480x + 75,600 = 167,760$$

Subtract 75,600: $\qquad 480x = 92,160$

Divide by 480: $\qquad\qquad x = 192 \,(\text{Adult tickets})$

$$280 - x = 88 \,(\text{Children's tickets})$$

Check

$$7.50x + 2.70(280 - x) = 1,677.60$$

$x = 192$: $\qquad 7.50(192) + 2.70(88) = 1,677.60$

$$1,440 + 237.60 = 1,677.60$$
$$1,677.60 = 1,677.60\checkmark$$

Answer

192 adult tickets and 88 children's tickets

Exercises

1a. A total of 210 tickets were sold at the junior prom. Members of the Student Organization paid $5, while nonmembers paid $11. If the total receipts amounted to $1,830, how many tickets were sold to Student Organization members?

1b. A total of 150 movie tickets were sold for a total of $744. If adult tickets sold for $7.20 and children's tickets sold for $3, how many tickets of each type were sold?

Example 2

Abraham Soto owns a men's clothing store. He sold out his entire stock of $64 and $51 jackets for a total of $3,000. If he sold 25 more $51 jackets than $64 jackets, how many jackets of each type did he sell?

Analysis

Let x = the number of $64 jackets.

Let $x + 25$ = the number of $51 jackets.

Type of Jacket	Number of Jackets, n	Price Per Jacket, p ($)	Total Price = $n \cdot p$ ($)
$64	x	64	$64x$
$51	$x + 25$	51	$51(x + 25)$

Work

The total price of the $64 jackets plus the total price of the $51 jackets equals $3,000.

$$64x + 51(x + 25) = 3,000$$

$$64x + 51x + 1,275 = 3,000$$

$$115x + 1,275 = 3,000$$

Subtract 1,275: $\qquad\qquad 115x = 1,725$

Divide by 115: $\qquad\qquad x = 15 \text{ ($64 jackets)}$

$$x + 25 = 40 \text{ ($51 jackets)}$$

Check

$$64x + 51(x + 25) = 3,000$$

$x = 15$: $\qquad 64(15) + 51(40) = 3,000$

$$960 + 2,040 = 3,000$$

$$3,000 = 3,000 \checkmark$$

Answer

Fifteen $64 jackets and forty $51 jackets

Exercises

2a. The Village Bookstore sells softcover and hardcover books. The average price of a softcover book is $14 and the average price of a hardcover book is $23. If, last week, the store sold 20 more softcover than hardcover books and had total sales of $3,610, how many books of each type did the store sell?

2b. Jeffrey is a tie salesperson. Yesterday, he sold 19 more solid ties than plaid ties. If each solid tie sold for $5 and each plaid tie sold for $6, and the total sales amounted to $810, how many ties of each type did Jeffrey sell?

2.3 Mixing Liquids

Chemists are always mixing stuff to come up with a new product. I have a friend who is a chemist. Before a major holiday he and his fellow chemists blend various mixtures together to produce a potent holiday drink! I advise you against doing the same; nevertheless, it is important to know how to mix various liquids.

The key to solving percent mixture problems is to add up the pure (unadulterated) parts of the mixtures and then to set the sum equal to the pure (unadulterated) part of the total mix. Or, to paraphrase Tom Cruise in the movie *Jerry Maguire* when he says "Show me the money," we'll insist "Show me pure liquids."

Example 1

Carey wants to mix a 12% solution of benzene with a 20% solution of benzene to make 160 ounces of a 15% solution. How many ounces of each type must Carey use?

Analysis

Carey will obtain a final mix of 160 ounces of a 15% solution. However, she doesn't know how many ounces of each of the initial ingredients to mix. Let's say she uses x ounces of the 12% solution. If the two solutions add up to 160 ounces, then the 20% solution weighs $160 - x$ ounces.

Let x = the total ounces of the 12% solution.

Let $160 - x$ = the total ounces of the 20% solution.

Kind	Percent Benzene	Total Ounces	Pure Benzene = Percent × Ounces
Solution 1	12	x	12%(x)
Solution 2	20	$160 - x$	20%($160 - x$)
Mixture	15	160	15%(160)

Work

The pure benzene is indicated in the last column on the right. The total benzene in solution 1 plus the total benzene in solution 2 equals the total benzene in the mixture.

$$12\%(x) + 20\%(160 - x) = 15\%(160)$$

Change % to decimals:

$$0.12(x) + 0.20(160 - x) = 0.15(160)$$

Multiply by 100:

$$12x + 20(160 - x) = 15(160)$$

$$12x + 3,200 - 20x = 2,400$$

$$-8x + 3,200 = 2,400$$

Subtract 3,200: $\quad -8x = -800$

Divide by –8: $\quad x = 100$ (ounces of the 12% solution)

$$160 - x = 60 \text{ (ounces of the 20\% solution)}$$

Check

$$12\%(x) + 20\%(160 - x) = 15\%(160)$$

$$0.12(x) + 0.20(160 - x) = 0.15(160)$$

$x = 100$: $\quad 0.12(100) + 0.20(60) = 0.15(160)$

$$12 + 12 = 24$$

$$24 = 24 \checkmark$$

Answer

100 ounces of the 12% solution, 60 ounces of the 20% solution.

1a.   The Bigelow Wine Company wants to blend a California wine with a 5% solution of alcohol with a French wine with a 9% solution of alcohol to get a mix of 200 gallons of a wine with a 7% alcoholic content. How many gallons of California wine and how many gallons of French wine must be blended?

1b.   A jeweler mixes an alloy of 16% gold with a second alloy of 28% gold. He wants to make a new alloy containing 32 ounces of 25% gold. How many ounces of each of the original alloys must he mix?

Exercises

Example 2

Jesus wants to change a 32-liter 7% solution of alcohol to a 4% solution by adding pure distilled water. How much pure distilled water should he add?

Analysis

Distilled water is pure water, which means that there is absolutely no (0%) alcohol in it.

Let x = the number of liters of pure distilled water added to the 4% solution to reduce it to a 7% solution of alcohol.

Kind	Percent Alcohol	Total Liters	Alcohol = Percent × Liters
Initial solution	7%	32	7%(32)
Added pure water	0%	x	0%(x)
Final mixture	4%	32 + x	4%(32 + x)

Work

The amount of alcohol in solution 1 plus the amount of alcohol in the added pure water equals the amount of alcohol in the mixture.

$$7\%(32) + 0\%(x) = 4\%(32 + x)$$

Change to decimals: $\quad 0.07(32) + 0(x) = 0.04(32 + x)$

Multiply by 100: $\quad\quad\quad 7(32) = 4(32 + x)$

$$224 = 128 + 4x$$

Subtract 128: $\quad\quad\quad\quad\quad 96 = 4x$

Divide by 4: $\quad\quad\quad\quad\quad\quad 24 = x$

Symmetric Property: $\quad\quad\quad\quad x = 24$

Check

$$7\%(32) + 0\%(x) = 4\%(32 + x)$$

$$0.07(32) + 0(x) = 0.04(32 + x)$$

$x = 24$: $\quad\quad\quad 0.07(32) + 0 = 0.04(56)$

$$2.24 = 2.24 \checkmark$$

Answer

24 liters of pure distilled water

2a. The Denver Chemical Company wants to increase the 10% peroxide content of its hydrogen peroxide product to a 28% solution by adding pure peroxide (100% peroxide). If Denver has 500 gallons of its original 10% peroxide, how much pure peroxide must it add to produce the new product?

2b. Macri wants to change 18 quarts of an 8% solution of ammonia to a 6% solution by adding pure water. How much pure water must she add?

Example 3

Of the 20 quarts of liquid in the radiator of a car, 55% is antifreeze. How much liquid must be drained off and replaced by an equal amount of a 75% solution of antifreeze to get a final mix of 70% antifreeze?

Analysis

Let x = the amount of liquid drained off and then just follow the antifreeze.

Work

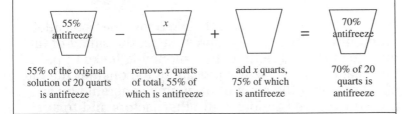

| 55% of the original solution of 20 quarts is antifreeze | remove x quarts of total, 55% of which is antifreeze | add x quarts, 75% of which is antifreeze | 70% of 20 quarts is antifreeze |

The antifreeze in the original solution (55% · 20) minus the antifreeze drained off (55% · x) plus the antifreeze added (75% · x) equals the antifreeze in the new solution (70% · 20).

$$55\%(20) - 55\%(x) + 75\%(x) = 70\%(20)$$

Change to decimals:

$$0.55(20) - 0.55(x) + 0.75(x) = 0.70(20)$$

Multiply by 100: $55(20) - 55(x) + 75(x) = 70(20)$

$$1{,}100 - 55x + 75x = 1{,}400$$

$$1{,}100 + 20x = 1{,}400$$

Subtract 1,100: $20x = 300$

Divide by 20: $x = 15$

Check

$$55\%(20) - 55\%(x) + 75\%(x) = 70\%(20)$$
$$0.55(20) - 0.55(x) + 0.75(x) = 0.70(20)$$

Multiply by 100: $\quad 55(20) - 55(x) + 75(x) = 70(20)$

$x = 15$: $\quad\quad\quad 55(20) - 55(15) + 75(15) = 1,400$

$$1,100 - 825 + 1,125 = 1,400$$
$$1,400 = 1,400 \checkmark$$

Answer

15 quarts

Exercises

3a. There are 20 quarts of a 40% solution of antifreeze in a car's radiator. How much of the solution must be drained off and replaced by an equal amount of a 60% solution to get a final mix of a 45% solution of antifreeze?

3b. Sean wants to change 40 liters of a 60% solution of liquid hydrogen by draining off some liters and replacing them with an equal amount of a 20% solution of liquid hydrogen. If he wants to get a final mix of 52% of liquid hydrogen, how many liters must he drain off?

2.4 Investment Problems

When people invest their money, they consider different factors such as interest rates, the safety of the investment, and the time their money is locked up. In this section, since we are just going to consider the rate of interest and the amount of money generated by an investment, we can forget all other factors and treat these questions as simple mixture problems.

Example 1

Milt has $7,000 to invest. If he invests part at 6% simple annual interest and part at 8% simple annual interest, he will get an annual return of $520. How much should Milt invest at each rate?

Analysis

Let x = the amount invested at 6%.

Let $7,000 - x$ = the amount invested at 8%.

Investment	Percent Interest	Principal ($)	Interest = Percent × Principal ($)
Investment 1	6%	x	$6\%(x)$
Investment 2	8%	$7{,}000 - x$	$8\%(7{,}000 - x)$

Work

The interest from the two investments adds up to $520.

$$6\%(x) + 8\%(7{,}000 - x) = 520$$

Change to decimals:

$$0.06(x) + 0.08(7{,}000 - x) = 520$$

Multiply by 100: $\quad 6(x) + 8(7{,}000 - x) = 52{,}000$

$$6x + 56{,}000 - 8x = 52{,}000$$

$$-2x + 56{,}000 = 52{,}000$$

Subtract 56,000: $\qquad\qquad -2x = -4{,}000$

Divide by −2: $\qquad\qquad\qquad x = 2{,}000 \,(\text{at } 6\%)$

$$7{,}000 - x = 5{,}000 \,(\text{at } 8\%)$$

Check

$$6\%(x) + 8\%(7{,}000 - x) = 520$$

$$0.06(x) + 0.08(7{,}000 - x) = 520$$

$x = 2{,}000$: $\quad 0.06(2{,}000) + 0.08(5{,}000) = 520$

$$120 + 400 = 520$$

$$520 = 520 \checkmark$$

Answer

$2,000 at 6% and $5,000 at 8%

In all cases, the interest rate is simple annual interest.

1a.   Mr. Johanssen has saved $9,000. If he invests a certain amount at 6% and the remainder at 7% and his total annual income from these two investments is $610, how much has he invested in each place?

1b.   Out of $11,000, Ms. Jefferson places some money in stocks at 5% and the remainder in bonds at 8%. Her total annual income is $730. How much did she invest in stocks and how much in bonds?

Example 2

Phuong buys $7,000 of French bonds and $12,000 of Mexican bonds. The Mexican bonds pay a simple annual interest rate 2% more than the French bonds. If the total annual return is $1,380, find the interest rate offered on each bond.

Analysis

Let x = the interest rate offered by the French bonds.

Let $x + 2$ = the interest offered by the Mexican bonds.

Investment	Percent Interest	Principal ($)	Interest = Percent × Principal ($)
French bonds	x	7,000	$x(7,000)$
Mexican bonds	$x + 2$	12,000	$(x + 2)(12,000)$

Work

The annual interest on the French bonds plus the annual interest on the Mexican bonds equals $1,380.

$$x(7,000) + (x + 2\%)(12,000) = 1,380$$

$2\% = 0.02$: $x(7,000) + (x + 0.02)(12,000) = 1,380$

$$7,000x + 12,000x + 240 = 1,380$$

$$19,000x + 240 = 1,380$$

Subtract 240: $19,000x = 1,140$

Divide by 19,000: $x = 0.06$
                      $= 6\%$ (French bonds)

                $x + 0.02 = 0.08$
                      $= 8\%$ (Mexican bonds)

Check

$$x(7,000) + (x + 2\%)(12,000) = 1,380$$
$$x(7,000) + (x + 0.02)(12,000) = 1,380$$
$x = 0.06$: $0.06(7,000) + 0.08(12,000) = 1,380$
$$420 + 960 = 1,380$$
$$1,380 = 1,380 \checkmark$$

Answer

French bonds offer 6%; Mexican bonds offer 8%.

2a. Hvlati invests $9,000 in state bonds and $12,000 in U.S. Treasury Bonds. He receives a total annual return of $1,410. If the state bonds pay an annual interest rate 4% more than the U.S. Treasury Bonds, what is the annual interest rate on each bond?

2b. The Risk Mutual Fund invests $150,000 in Corporation A and $250,000 in Corporation B. If the total annual return for both investments is $22,500 and Corporation A pays 1% more interest than Corporation B, what are the interest rates for both investments?

Exercises

Coin and stamp problems are really just a subspecies of mixture problems. We're just mixing different denominations of coins or stamps rather than liquids or solids.

2.5 Coin and Stamp Problems

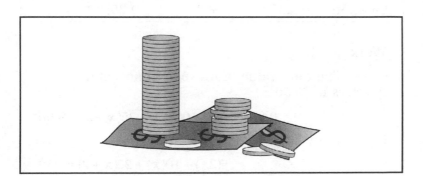

Example 1

How many cents are there in 6 dimes?

$$6 \times 10 = 60 \text{ cents}$$

Change 8 quarters to cents.

$$8 \times 25 = 200 \text{ cents}$$

How much is $(3y + 4)$ dimes, in cents?

$$(3y + 4)10 = 30y + 40 \text{ cents}$$

Exercises

1a. How many cents are there in 24 dimes?

1b. Change 18 quarters to cents.

1c. Change $2x + 3$ dimes to cents.

Example 2

Hilda has saved $9.50 in nickels, dimes, and quarters. She has twice as many nickels as dimes and two more quarters than dimes. How many coins of each kind does she have?

Analysis

Let x = the number of dimes.

Let $2x$ = the number of nickels.

Let $x + 2$ = the number of quarters.

Coin	Value ($)	Number	Total Value = Value × Number ($)
Nickels	0.05	$2x$	$0.05(2x)$
Dimes	0.10	x	$0.10(x)$
Quarters	0.25	$x + 2$	$0.25(x + 2)$

Work

The total value of nickels, dimes, and quarters is $9.50.

$$0.05(2x) + 0.10(x) + 0.25(x + 2) = 9.50$$

Multiply by 100:

$$5(2x) + 10(x) + 25(x + 2) = 950$$

$$10x + 10x + 25x + 50 = 950$$

$$45x + 50 = 950$$

Subtract 50:	$45x = 900$
Divide by 45:	$x = 20\,(\text{dimes})$
	$2x = 40\,(\text{nickels})$
	$x + 2 = 22\,(\text{quarters})$

Check

$$0.05(2x) + 0.10(x) + 0.25(x + 2) = 9.50$$

$$0.05(40) + 0.10(20) + 0.25(22) = 9.50$$

$$2.00 + 2.00 + 5.50 = 9.50$$

$$9.50 = 9.50\,\checkmark$$

Answer

20 dimes, 40 nickels, and 22 quarters

Exercises

2a.   Louis has $5.90 in nickels and dimes. If he has 22 more nickels than dimes, determine the number of each type of coin.

2b.   Dino's Diner has $29.50 in nickels and dimes. There are 115 more dimes than nickels. Find the number of nickels and dimes.

Example 3

José has some nickels, dimes, and quarters totaling $5.30. He has 37 coins altogether and he has four more nickels than dimes. How many coins of each type does he have?

Analysis

Let x = the number of dimes.

Let $x + 4$ = the number of nickels.

Let the number of quarters
$= 37 - \{(x) + (x + 4)\} = 37 - (x + x + 4)$
$= 37 - (2x + 4)$.

Coin	Value ($)	Number	Total Value = Value × Number ($)
5¢	0.05	$x + 4$	$0.05(x + 4)$
10¢	0.10	x	$0.10(x)$
25¢	0.25	$37 - (2x + 4)$	$0.25[37 - (2x + 4)]$

The total collection is valued at $5.30.

$$0.10x + 0.05(x+4) + 0.25[37 - (2x+4)] = 5.30$$

Multiply by 100:

$$10x + 5(x+4) + 25[37 - (2x+4)] = 530$$

$$10x + 5x + 20 + 25[37 - (2x+4)] = 530$$

$$15x + 20 + 25[(33 - 2x)] = 530$$

$$15x + 20 + 825 - 50x = 530$$

$$-35x + 845 = 530$$

Subtract 845: $\qquad -35x = -315$

Divide by −35: $\qquad x = 9 \, (\text{dimes})$

$$x + 4 = 13 \, (\text{nickels})$$

$$37 - (2x+4) = 37 - (18+4) = 37 - 22 = 15 \, (\text{quarters})$$

Check

$$0.10x + 0.05(x+4) + 0.25[37 - (2x+4)] = 5.30$$

$x = 9$: $\qquad 0.10(9) + 0.05(13) + 0.25(15) = 5.30$

$$0.90 + 0.65 + 3.75 = 530$$

$$5.30 = 5.30 \checkmark$$

Answer

9 dimes, 13 nickels, and 15 quarters

Exercises

3a. Rubin bought some candy for $1.85. If he paid in dimes and quarters and there were 11 coins altogether, find the number of each type of coin.

3b. Maggie has $4.75 in nickels, dimes, and quarters in her purse. If she has 31 coins altogether and there are twice as many dimes as nickels, how many coins of each type does she have?

Example 4

Valerie buys 8¢, 20¢, and 33¢ stamps. She buys half as many 8¢ stamps as 20¢ stamps and four fewer 33¢ stamps than 8¢ stamps. The entire purchase costs $6.78. How many stamps of each kind does Valerie purchase?

Analysis

Let x = the number of 20¢ stamps.

Let $\frac{1}{2}x$ = the number of 8¢ stamps.

Let $\frac{1}{2}x - 4$ = the number of 33¢ stamps.

Stamp	Value ($)	Number	Total Value = Value × Number ($)
8¢	0.08	$\frac{1}{2}x$	$0.08(\frac{1}{2}x)$
20¢	0.20	x	$0.20(x)$
33¢	0.33	$\frac{1}{2}x - 4$	$0.33(\frac{1}{2}x - 4)$

The total value of the purchase is $6.78.

$$0.20(x) + 0.08\left(\frac{1}{2}x\right) + 0.33\left(\frac{1}{2}x - 4\right) = 6.78$$

Multiply by 100:

$$20(x) + 8\left(\frac{1}{2}x\right) + 33\left(\frac{1}{2}x - 4\right) = 678$$

$$20x + 4x + \frac{33x}{2} - 132 = 678$$

Multiply by 2: $\quad 40x + 8x + 33x - 264 = 1{,}356$

$$81x - 264 = 1{,}356$$

Add 264: $\qquad\qquad\quad 81x = 1{,}620$

Divide by 81: $\qquad\qquad\quad x = 20$

$$\frac{1}{2}x = 10$$

$$\frac{1}{2}x - 4 = 6$$

Check

$$0.20(x) + 0.08\left(\frac{1}{2}x\right) + 0.33\left(\frac{1}{2}x - 4\right) = 6.78$$

$$0.20(20) + 0.08(10) + 0.33(6) = 6.78$$

$$4.00 + 0.80 + 1.98 = 6.78$$

$$6.78 = 6.78\checkmark$$

Answer

Twenty 20¢ stamps, ten 8¢ stamps, six 33¢ stamps

4a. Harry bought some $0.20, $0.30, and $0.34 stamps. He purchased half as many $0.30 stamps as $0.20 stamps and five more $0.34 stamps than $0.30 stamps. If the total purchase was $5.86, how many stamps of each type did he buy?

4b. Mai has nickels, dimes, and quarters in her purse. She has two thirds as many dimes as nickels and half as many quarters as nickels. If the total value of all of her coins is $2.90, how many coins of each type does she have?

Test

1. Malika's Candy sells for $0.43 an ounce. What is the cost of 26 pounds of candy?

2. Rico mixes 15 pounds of chocolate chip cookies at $4.64 per pound with 25 pounds of jelly cookies at $3.36 per pound. Determine the price of 1 pound of the new mixture.

3. A mixture of 24 pounds of nails selling for $2.50 per pound contains five times as many nails selling at $2.40 a pound as nails selling at $3.00 a pound. How many pounds of nails at $3.00 per pound are in the mixture?

4. The local civic club held a fundraiser. Members were charged $4.50 a ticket, and nonmembers were charged $8.25. If the total receipts amounted to $1,222.50 and 180 people attended, how many were members and how many were nonmembers?

5. One week, Crandall's Computer Store sold computers at $1,180 per unit and hard drives at $125 per unit. If the store sold eight more computer units than hard drives and the total sales amounted to $27,710, how many of each did the store sell?

6. Malcolm wants to mix a 4% peroxide solution with an 8% solution of peroxide to get a 260-liter 5.4% solution of peroxide. How many liters of each type must he mix?

7. Rosa wants to dilute an 80-ounce 9% solution of hydrochloric acid to a 3% solution by adding pure distilled water. How much water should she add?

8. Jang has $9,000 to invest. He places some money in bonds of the Sterling Corporation paying 4% simple annual interest and the remainder in bonds of the HGT Corporation paying 7% simple annual interest. If the total annual income is $555, how much did he place in each corporation?

9. Wladislav has $8.50 in nickels, dimes, and quarters. If he has four more dimes than nickels and

three times as many quarters as nickels, how many coins of each type does he have?

10.   The capacity of a car radiator is 20 quarts. The antifreeze in the radiator is 30% and Jean wishes to drain some off and replace it with an equal amount of a 50% solution. How much should she drain off to have a 45% solution?

11.   Helen has $2.35 in nickels, dimes, and quarters. If she has 18 coins altogether and she has two more nickels than dimes, how many coins of each type does she have?

12.   Joe buys 10¢, 20¢, and 34¢ stamps. If his total purchase amounts to $9.78 and he buys twice as many 20¢ stamps as 10¢ stamps and nine more 34¢ stamps than 10¢ stamps, how many stamps of each type does he buy?

13.   The Denver Chemical Company wants to increase its 10% peroxide to a 28% solution by adding 100% peroxide. If Denver has 500 gallons of its original 10% peroxide, how much pure peroxide must it add to obtain the new mixture?

14.   Joe Calendra saved $8,000. He placed $2,000 in a regular savings account at $5\frac{1}{2}$% simple annual interest and the remainder in a money market account at $6\frac{1}{4}$% simple annual interest. Find his total annual interest.

15.   Kim Powell manufactures metal nails. She manufactures two different types of nails and packages 55 of them together in one envelope. If nail A weighs 0.1 ounce, nail B weighs 0.15 ounce, and the entire package weighs 6.75 ounces, how many nails of each type does she package?

16.   Jerry's Candy Corporation sells a wholesale mix of mints and chocolates. There are 6 pounds more of chocolates than there are of mints. Mints sell for $4.60 a pound, chocolates sell for $6.90 a pound, and the mix sells for $5.98 a pound. How many pounds each of chocolates and mints are there in the mix?

17.   The Zap Electronic Store had a sale on calculators. It sold some calculators at $15.75 each and some other calculators at $18.50 each. If the store sold five more of the $15.75 calculators than it did of the $18.50 calculators and the total sales amounted to$1,757, how many of each type of calculator were sold?

18.   Kendra wants to mix a 6% solution of sulfuric acid with a 9% solution of sulfuric acid to obtain an 8% mixture. There are 12 quarts more of the 9% solution than of the 6% solution. How many quarts are there in the mixture?

19. Mr. Bilbao invests $15,000 in a real estate syndicate and $19,000 in a mortgage. If the mortgage pays an annual interest rate that is 2% less than the real estate syndicate and the total annual return on both investments is $2,680, what are the interest rates for each investment?

20. How many cents are there in 15 dimes and 6 quarters?

3

Perimeters and Circumferences

By definition, the **perimeter** of a two-dimensional figure is defined as the sum of the sides of that figure.

A **polygon** is a two-dimensional (or plane) figure enclosed by three or more straight lines. The polygons we encounter most often are

3.1 Polygons

triangle	3 sides
equilateral triangle	3 equal sides + 3 equal angles
quadrilateral	4 sides
rhombus	4 equal sides
square	4 equal sides + 4 right angles

rectangle	opposite sides equal + 4 right angles
pentagon	5 sides
hexagon	6 sides
octagon	8 sides
decagon	10 sides
"regular" polygon	all sides equal and all angles equal

Example 1

Marion wants to build a fence in the shape of a hexagon around her property. The sides of the fence are 49.4 feet, 34.8 feet, 53.6 feet, 65.8 feet, 72.5 feet, and 48.5 feet. If fencing costs $5.36 per foot, find the cost for the complete fence. Round off to the nearest cent.

Analysis

First, find the perimeter by adding up the six sides. Then, multiply the perimeter by $5.36, the cost of fencing per foot.

Work

$$49.4 + 34.8 + 53.6 + 65.8 + 72.5 + 48.5 = 324.6$$

Multiply by $5.36: $324.6 \times \$5.36 = \$1,739.856$

$$\approx \$1,739.86$$

Answer

$1,739.86

Exercises

1a. The sides of a pentagon-shaped area are 49.2 feet, 56.4 feet, 62.7 feet, 47.3 feet, and 65.4 feet. If a fence costs $3.93 per foot, what is the cost of a fence for the entire perimeter?

1b. Jose wants to make an octagon-shaped art-deco piece and trim it with silver. If silver costs $0.09 per inch and the sides of the piece are 12 inches, 14 inches, 13 inches, 16 inches, 19 inches, 21 inches, 23 inches, and 24 inches, find the cost of the trim.

Example 2

The perimeter of a pentagon is 64 inches. If four sides are 11 inches, 17 inches, 13 inches, and 9 inches, find the fifth side.

Analysis

Let x = the missing fifth side.

A pentagon has five sides, so if we're given four of those sides, just add them up and subtract from the perimeter.

Work

$$x + 11 + 17 + 13 + 9 = 64$$
$$x + 50 = 64$$

Subtract 50:
$$x = 14$$

Check

$$x + 11 + 17 + 13 + 9 = 64$$
$$14 + 11 + 17 + 13 + 9 = 64$$
$$64 = 64 \checkmark$$

Answer

14

Exercises

2a. The perimeter of a triangle is 34.8. If two sides are 19.4 and 8.5, find the third side.

2b. A quadrilateral has three sides measuring 14.5, 8.9, and 10.2. If the perimeter is 45.6, find the fourth side.

Example 3

The sides of an octagon are represented by x, $x + 2$, $2x$, $2x + 2$, $2x + 3$, $3x$, $3x + 1$, and $4x$. If the perimeter is 98, find each side of the octagon.

Analysis

Just add up the eight given sides and set the total equal to the perimeter, 98.

Work

$$x + (x+2) + (2x) + (2x+2) + (2x+3) + (3x)$$
$$+ (3x+1) + (4x) = 98$$
$$18x + 8 = 98$$

Subtract 8: $\qquad 18x = 90$

Divide by 18: $\qquad x = 5$

$$x + 2 = 7$$
$$2x = 10$$
$$2x + 2 = 12$$
$$2x + 3 = 13$$
$$3x = 15$$
$$3x + 1 = 16$$
$$\underline{4x = 20}$$

Check $\qquad\qquad\qquad 98 \checkmark$

Answer

5, 7, 10, 12, 13, 15, 16, 20

Exercises

3a. The sides of a pentagon are represented by x, $x + 4$, $2x + 1$, $2x + 2$, and $3x$. If the perimeter is 52, find each side.

3b. The perimeter of a hexagon is 87. If the sides are represented by x, $x + 4$, $2x$, $2x + 1$, $2x + 5$, and $3x$, find each side.

A **regular polygon** is a polygon with equal sides and equal angles.

Example 4

A regular pentagon has a perimeter of 41.5. Find one side of the pentagon.

Analysis

A regular pentagon has five equal sides, so just divide 41.5 by 5.

Work
$$\frac{41.5}{5} = 8.3$$

Answer

8.3

4a.   The perimeter of a regular hexagon is 28.8. Find one side.

4b.   If the perimeter of a regular decagon is 33.75, find one side.

Example 5

Each side of a regular octagon is represented by $5x - 6$. If the perimeter is 72, find x.

Analysis

Let P = the perimeter of the regular octagon.

Let s = one side of the regular octagon.

A regular octagon has eight equal sides, so just multiply 8 by the length of one side, s, and set the resulting equation equal to 72.

Work

$$P = 8s$$

$P = 72$, $s = 5x - 6$: $72 = 8(5x - 6)$

$72 = 40x - 48$

Add 48: $120 = 40x$

Divide by 40: $3 = x$

Symmetric Property: $x = 3$

$$5x - 6 = 5(3) - 6 = 15 - 6 = 9$$

Check

$$P = 8s$$

$P = 72$, $s = 8$: $72 = 8(9)$

$72 = 72$ ✓

Answer

3

5a.   Each side of a regular quadrilateral is represented by $3x + 2$. If the perimeter is 44, find x.

5b.   The perimeter of an equilateral triangle is 87. If each side is represented by $5x - 1$, find x.

Example 6

A side of a regular hexagon is 3 less than twice the length of a side of a regular decagon. If the perimeter of the hexagon equals the perimeter of the decagon, find a side of the hexagon.

Analysis

A hexagon has 6 sides, while a decagon has 10 sides.

Let x = a side of a regular decagon.

Let $2x - 3$ = a side of a regular hexagon.

The perimeter of a regular decagon = $10x$.

The perimeter of a regular hexagon = $6(2x - 3)$.

Work

The perimeters are equal:

$$6(2x - 3) = 10x$$
$$12x - 18 = 10x$$

Subtract $10x$: $\quad 2x - 18 = 0$

Add 18: $\quad\quad\quad 2x = 18$

Divide by 2: $\quad\quad x = 9$ (side of the regular decagon)

$\quad\quad\quad\quad 2x - 3 = 15$ (side of the regular hexagon)

Check

The perimeter of a regular hexagon equals the perimeter of a regular decagon.

$x = 9$, $2x - 3 = 15$: $\quad 6(15) = 10(9)$

$$90 = 90 \checkmark$$

Answer

15

6a. Each side of an equilateral triangle is 6 less than four times the side of a regular decagon. If the perimeters of both figures are equal, find a side of the triangle.

6b. A side in a rhombus is 3 more than twice the side of a regular hexagon. If the perimeter of the rhombus is twice the perimeter of the regular hexagon, find a side of the regular hexagon.

A **rectangle** is a quadrilateral with four right angles and with both pairs of opposite sides equal. A rectangle is actually a particular kind of square. The perimeter equals the sum of the sides:

$$P = l + l + w + w = 2l + 2w$$

The perimeter equals 2 lengths plus 2 widths.

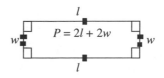

Example 1

A glass window is 5 feet 4 inches long and 3 feet 7 inches wide. In preparation for a hurricane, we want to tape the *inside perimeter* of the glass window. How much tape do we need, in feet and inches?

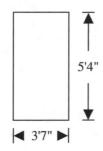

5'4"

3'7"

Analysis

Let P = the perimeter of the rectangle.

Let l = the length of the rectangle.

Let w = the width of the rectangle.

Perimeter = 2 × length + 2 × width.

1 foot = 12 inches.

Work

$$P = 2l + 2w$$
$$P = 2 \times 5'4'' + 2 \times 3'7''$$
$$P = 10'8'' + 6'14''$$
$$P = 16'22''$$

1 foot = 12 inches, so 22 inches = 1 foot + 10 inches.

$$P = 16' + 1' + 10'' = 17'10''$$

Answer

17 ft 10 in.

1a. A bedroom is 16′7″ long by 12′3″ wide. Find the perimeter of the room, in feet and inches.

1b. A living room is 37′9″ long by 26′5″ wide. Find the perimeter, in feet and inches.

Example 2

Margie wants to frame a rectangular painting. If the perimeter is 13 ft 4 in. and the width is 2 ft 6 in., find the length of the painting.

$$w = 2'6''$$

Analysis

Let P = the perimeter of the rectangle.

Let l = the length of the rectangle.

Let w = the width of the rectangle.

The perimeter equals the sum of the four sides, so $P = 2l + 2w$. Substitute values for P and w into the formula. To simplify things, it's probably easier to change all the figures to inches.

Work

$$w = 2 \text{ feet } 6 \text{ inches}$$
$$= 2 \times 12'' + 6'' = 30''$$
$$P = 13 \text{ feet } 4 \text{ inches}$$
$$= 13 \times 12 + 4 = 160''$$
$$P = 2l + 2w$$

$P = 160,\ w = 30$:
$$160 = 2l + 2(30)$$
$$160 = 2l + 60$$

Subtract 60:
$$100 = 2l$$

Divide by 2:
$$50 = l$$

Check

$$P = 2l + 2w$$

$l = 50$, $w = 30$,
$P = 160$:

$$160 = 2(50) + 2(30)$$

$$160 = 100 + 60$$

$$160 = 160 \checkmark$$

Answer

length = 50 inches or 4 ft 2 in.

2a.   A painting has a perimeter of 7′8″. If the painting is 2′5″ in length, find its width.

2b.   A garden is in the shape of a rectangle. If the perimeter of the garden is 102′4″ and its width is 22′8″, find its length.

Example 3

The length of a rectangle is 3 inches more than two times the width. If the perimeter is 36 inches, find the length and width of the rectangle.

$$l = 2w + 3$$

$w \quad \boxed{\quad P = 2l + 2w \quad} \quad w$

$$l = 2w + 3$$

Analysis

Let P = the perimeter of the rectangle.

Let l = the length of the rectangle.

Let w = the width of the rectangle.

$P = 2l + 2w$

The length is 3 inches more than twice the width: $l = 2w + 3$.

Work

$$P = 2l + 2w$$

$P = 36$, $l = 2w + 3$: $\quad 36 = 2(2w + 3) + 2w$

$$36 = 4w + 6 + 2w$$

$$36 = 6w + 6$$

Subtract 6: $\qquad 30 = 6w$

Divide by 6: $\qquad 5 = w$

Symmetric Property: $\qquad w = 5$

$$l = 2w + 3$$

$$l = 2(5) + 3 = 13$$

Check

$$P = 2l + 2w$$

$P = 36,\ w = 5,\ l = 13:\qquad 36 = 2(13) + 2(5)$

$$36 = 26 + 10$$

$$36 = 36 \checkmark$$

Answer

length = 13″, width = 5″

Exercises

3a. The length of a rectangular room is 4 feet less than twice the width. If the perimeter of the room is 40 feet, find the dimensions.

3b. The perimeter of a rectangle is 50. If the length is 5 more than three times the width, find the dimensions of the rectangle.

3.3 Squares

A square is a regular polygon with four equal sides and four right angles. Since, by definition, all the sides of a square are equal, let's call each side s. The perimeter, P, equals 4 times the length of one side, s:

$$P = 4s$$

Example 1

Joe is up at bat. The baseball diamond is actually a skewed square, with each side equal to 90 feet.

  a. Find the perimeter.

  b. If Joe runs at the rate of 18 feet/sec, how long would it take for him to run around all the bases?

Analysis

    a.   Let P = the perimeter of the square.

         Let s = a side of the square.

         Use the formula $P = 4s$ and substitute $s = 90$.

    b.   To find Joe's time around the bases, divide the perimeter by Joe's speed, 18 feet/sec.

Work

    a. $P = 4s$

$s = 90$: $P = 4(90) = 360$ feet

    b. $\dfrac{\text{Perimeter}}{\text{Joe's speed}} = \dfrac{360\,\text{ft}}{18\,\text{ft/sec}} = 20\,\text{sec}$

Answers

    a.   360 feet

    b.   20 seconds

1a.   A park is laid out in the shape of a square and each side is 935 feet.
    1.   Find the perimeter.
    2.   Walking at the rate of 440 feet/min, how long would it take to walk around the perimeter of the park?

1b.   A square block is 567 feet on each side.
    1.   What is the perimeter of the block?
    2.   If Danielle walks at the rate of 10.08 feet/sec, how long will it take for her to walk around the block? Leave your answer in seconds.

Exercises

Example 2

Kanequa wants to install molding around the entire base of a living room. Each side of the square living room is 16 feet.

   a. How many feet of molding does she need for the entire perimeter?

   b. If the molding costs $1.26 per foot, how much is the cost of molding for the entire perimeter?

Analysis

Let P = the perimeter of the square.

Let s = a side of the square.

Kanequa has a square living room, so she has to use the formula $P = 4s$. Once she determines the perimeter of the living room, she has to multiply that number by $1.26, the cost per foot of molding.

Work

   a. $P = 4s$

$s = 16$: $P = 4(16)$

          $P = 64$

   b. Total cost $= (64) \times (\$1.26) = \$80.64$

Answer

   a. Perimeter = 64 feet

   b. Total cost = $80.64

Exercises

2a. Jason's bedroom is in the shape of a square, and each wall is 14 feet. Jason wants to place molding around the room.
   1. Find the perimeter of the room.
   2. Find the cost of molding for the perimeter if each foot of molding costs $0.29.

2b. Malcolm wants to plant bushes around the perimeter of his property. The property is in the shape of a square and each side is 76 feet.
   1. Find the perimeter of the property.
   2. If bushes cost $1.34 a foot, find the cost of planting bushes around the perimeter of the property.

Example 3

Mr. and Mrs. Robinson want to put up a fence around their property. The property is in the shape of a square and the perimeter is 260 feet.

   a. Find one side of the property.

   b. If the cost of fencing the entire property is $899.60, how much is each foot of fence?

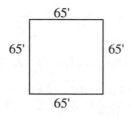

Analysis

   a. Let P = the perimeter of the square.

       Let s = a side of the square.

       Use the formula $P = 4s$ to find one side.

   b. Divide $899.60 by the perimeter, 260.

Work

     a. $P = 4s$

$P = 260$: $260 = 4s$

Divide by 4: $65 = s$

Symmetric property: $s = 65\,\text{ft}$

     b.

$$\text{Cost per foot} = \frac{\text{Total cost of perimeter fencing}}{\text{Number of feet in perimeter}}$$

$$= \frac{\$899.60}{260} = \$3.46$$

Answers

   a. 65 feet

   b. $3.46 per foot

3a. Hilda wants to tile the perimeter of her pool. The pool is in the shape of a square and the perimeter is 96 feet.
 1. Find one side of the pool.
 2. If the cost of tiling the perimeter of the pool is $353.28, find the cost of 1 foot of tile.

3b. Ms. White wants to put up a chain-linked fence around her property. The property is in the shape of a square and the perimeter is 492 feet.
 1. Find one side of the property.
 2. If the cost of putting up the fence is $1,348.08, how much is the cost of 1 foot of fence?

Example 4

If a side of a square is represented by $x + 3$ and the perimeter is 64 inches, find x.

$$x + 3$$

$x + 3$	$P = 64$	$x + 3$

$$x + 3$$

Analysis

Let P = the perimeter of the square.

Let $x + 3$ = a side of the square.

Use the formula $P = 4s$ and substitute for P.

Work

$$P = 4s$$

$P = 64$, $s = x + 3$: $\qquad 64 = 4(x + 3)$

$$64 = 4x + 12$$

Subtract 12: $\qquad 52 = 4x$

Divide by 4: $\qquad 13 = x$

Symmetric Property: $\qquad x = 13$

Answer

13

4a. Each side of a square is represented by $x + 6$, and the perimeter is 32. Find x.

4b. The perimeter of a square is 44. If each side of the square is represented by $x - 3$, find x.

A **triangle** is a polygon with three sides. The perimeter of a triangle equals the sum of all its sides:

Perimeter = $Side_1 + Side_2 + Side_3$: $P = s_1 + s_2 + s_3$

Example 1

Iris wants to install a fence around a triangular play area. If the fence costs \$4.12 a foot and the sides of the triangle are 25.3 ft, 19.5 ft, and 26.5 ft, what is the cost of the fence? Round off to the nearest cent.

Analysis

Let P = the perimeter of the triangle.

Let s_1 = the first side of the triangle.

Let s_2 = the second side of the triangle.

Let s_3 = the third side of the triangle.

Iris must first find the perimeter. Then she has to multiply the number of feet in the perimeter by \$4.12, the cost of the fence.

Work

Perimeter =
$Side_1 + Side_2 + Side_3$: $P = s_1 + s_2 + s_3$

$s_1 = 25.3$, $s_2 = 19.5$, $s_3 = 26.5$: $P = 25.3 + 19.5 + 26.5$

$P = 71.3$

Multiply the number of feet in the perimeter, 71.3, by the cost of the fence, \$4.12 per foot: $71.3 \times \$4.12 = \$293.756$

$\approx \$293.76$

Answer

\$293.76

1a.   A triangular-shaped piece of land has sides of 45.7 feet, 63.8 feet, and 79.5 feet. If Jamal wants to enclose the area with a chain fence and the fence costs $2.26 per foot, what is the total cost of the fence?

1b.   Ms. Almanzar wants to tape the perimeter of a triangular play area in the school gymnasium. If the sides of the triangle are 45 feet 5 inches, 52 feet 2 inches, and 39 feet 5 inches and the tape costs $0.02 a foot, find the cost of taping the perimeter of the triangular area.

An **isosceles triangle** is a triangle with at least two equal sides.

An **equilateral triangle** is a triangle with three equal sides.

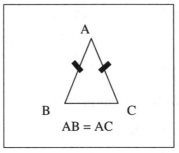

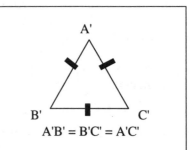

Example 2

The base of an isosceles triangle is 12.6 inches. If its perimeter is 32 inches, find one of the equal sides.

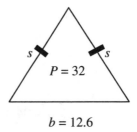

Analysis

Let s = each of the two equal sides of the isosceles triangle.

Let b = the base of the triangle.

Let P = the perimeter of the triangle.

The perimeter of an isosceles triangle equals the sum of the two equal sides plus the base:

$$P = 2s + b$$

Work

$$P = 2s + b$$

$P = 32, b = 12.6$: $32 = 2s + 12.6$

Subtract 12.6: $19.4 = 2s$

Divide by 2: $9.7 = s$

Symmetric Property: $s = 9.7$ (side)

Check

$$P = 2s + b$$
$$32 = 2(9.7) + 12.6$$
$$32 = 32 \checkmark$$

Answer

9.7″

2a. The base of an isosceles triangle is 14.5 and its perimeter is 35. Find one of the equal sides.

2b. The perimeter of an isosceles triangle is 42 and the base is 9.3. Find one of the equal sides.

Exercises

Example 3

One of the equal sides of an isosceles triangle is 6. If the perimeter is 28.1, find the base.

Analysis

Let P = the perimeter of the triangle.

Let s = each of the equal sides of the isosceles triangle.

Let b = the base of the isosceles triangle.

The perimeter is equal to the sum of the two equal sides and the base.

Work

$$P = 2s + b$$

$P = 28.1,\ s = 6:$ $28.1 = 2(6) + b$

                          $28.1 = 12 + b$

Subtract 12: $16.1 = b$

Symmetric Property: $b = 16.1\,(\text{base})$

Answer

　　16.1

Exercises

3a.   One of the equal sides of an isosceles triangle is 8 and the perimeter is 35. Find the base.

3b.   The perimeter of an isosceles triangle is 42. If one of the equal sides is 11.2, find the base.

Example 4

Each of the equal sides of an isosceles triangle is three more than the base. If the perimeter is 51, find each side of the triangle.

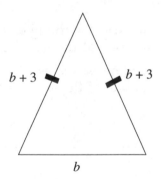

Analysis

Let P = the perimeter of the isosceles triangle.

Let b = the base of the triangle.

Let $b + 3$ = each of the equal sides of the triangle.

Work

$$P = 2(b + 3) + b$$

$P = 51:$ $51 = 2b + 6 + b$

$$51 = 3b + 6$$

Subtract 6: $\qquad\qquad 45 = 3b$

Divide by 3: $\qquad\qquad 15 = b$

Symmetric Property: $\qquad b = 15$ (base)

$$b + 3 = 18 \text{ (side)}$$

Check

$$51 = 2(b + 3) + b$$

$b = 15$: $\qquad 51 = 2(18) + 15$

$$51 = 36 + 15$$

$$51 = 51 \checkmark$$

Answer

Base = 15; each of the equal sides = 18

4a.   If each of the equal sides of an isosceles triangle is 5 more than the base and the perimeter is 34, find each side.

4b.   If the perimeter of an isosceles triangle is 39 and each of the equal sides is 3 more than the base, find each side of the triangle.

Example 5

The perimeter of an isosceles triangle is 33. If the base is 3 less than two times one of the equal sides, find each side of the triangle.

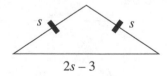

Analysis

Let $P =$ the perimeter of the isosceles triangle.

Let $s =$ one of the equal sides of the isosceles triangle.

The base is three less than two times one of the equal sides:

Let $2s - 3 =$ the base of the isosceles triangle.

Work

$$P = 2s + (2s - 3)$$

$P = 33$: $\qquad\qquad 33 = 2s + 2s - 3$

$\qquad\qquad\qquad\qquad 33 = 4s - 3$

Add 3: $\qquad\qquad\quad 36 = 4s$

Divide by 4: $\qquad\quad\; 9 = s$

Symmetric Property: $\quad s = 9\,\text{(side)}$

$\qquad\qquad\qquad 2s - 3 = 15\,\text{(base)}$

Check

$$P = 2s + (2s - 3)$$

$S = 9,\, b = 15,\, P = 33$: $\quad 33 = 2(9) + 15$

$\qquad\qquad\qquad\qquad 33 = 18 + 15$

$\qquad\qquad\qquad\qquad 33 = 33\;\checkmark$

Answer

Base = 15; each of the equal sides = 9

Exercises

5a.   The base of an isosceles triangle is 2 more than one of the equal sides. If the perimeter is 41, find each side of the triangle.

5b.   The perimeter of an isosceles triangle is 55. If the base is 2 less than each of the equal sides, find each side of the triangle.

Example 6

The perimeter of an equilateral triangle is 231.3. Find one side.

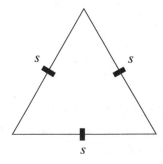

Analysis

Let $P =$ the perimeter of the equilateral triangle.

Let s = each side of the equilateral triangle.

Since all the sides of an equilateral triangle are equal, all we have to do is to divide the perimeter by 3.

Work

$$P = 3s$$
$$231.3 = 3s$$

Divide by 3: $77.1 = s$

Symmetric Property: $s = 77.1$

Answer

77.1

6a.   The perimeter of an equilateral triangle is 44.4. Find one side.

6b.   Find one side of an equilateral triangle if the perimeter is 77.1.

Exercises

Example 7

One side of an equilateral triangle is 34.8. Find the perimeter.

Analysis

Let P = the perimeter of the equilateral triangle.

Let s = each side of the equilateral triangle.

Since we know one side of the triangle and all the sides of an equilateral triangle are equal, just multiply that one side by three.

Work

$$P = 3s$$

$s = 34.8$: $P = 3(34.8) = 104.4$

Answer

104.4

7a.   One side of an equilateral triangle is 4.84. Find the perimeter.

7b.   Find the perimeter of an equilateral triangle if each side is 18.4.

Exercises

Example 8

A side of an equilateral triangle is represented by $2x + 3$. If the perimeter equals 45, find x.

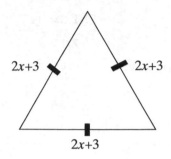

Analysis

Let P = the perimeter of the equilateral triangle.

Let $2x + 3$ = each side of the equilateral triangle.

Since an equilateral triangle, by definition, has three equal sides, the perimeter equals the sum of the three equal sides.

Work

$$P = 3(2x + 3)$$

$P = 45$:
$$45 = 3(2x + 3)$$
$$45 = 6x + 9$$

Subtract 9:
$$36 = 6x$$

Divide by 6:
$$6 = x$$

Symmetric Property:
$$x = 6$$

Substitute 6 for x in one side: $2x + 3 = 15$ (side)

Check

$$P = 3(2x + 3)$$
$$45 = 3(15)$$
$$45 = 45 \checkmark$$

Answer

$$x = 6$$

8a. The perimeter of an equilateral triangle is 48. If each side is represented by $3x - 2$, find x.

8b. One side of an equilateral triangle is represented by $2x - 3$. If the perimeter is 33, find x.

A **parallelogram** is a quadrilateral with opposite sides parallel (∥). The opposite sides and angles are congruent.

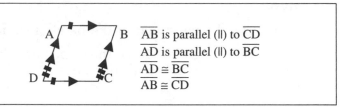

$\overline{AB}$ is parallel (∥) to $\overline{CD}$
$\overline{AD}$ is parallel (∥) to $\overline{BC}$
$\overline{AD} \cong \overline{BC}$
$\overline{AB} \cong \overline{CD}$

A **trapezoid** is a quadrilateral with two parallel sides and two nonparallel sides. In an **isosceles trapezoid**, the two nonparallel sides are congruent.

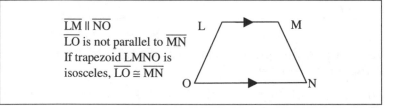

$\overline{LM} \parallel \overline{NO}$
$\overline{LO}$ is not parallel to $\overline{MN}$
If trapezoid LMNO is isosceles, $\overline{LO} \cong \overline{MN}$

A **rhombus** is a parallelogram with all four sides congruent and opposite sides parallel.

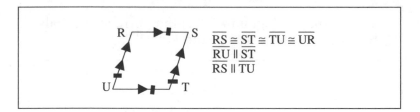

$\overline{RS} \cong \overline{ST} \cong \overline{TU} \cong \overline{UR}$
$\overline{RU} \parallel \overline{ST}$
$\overline{RS} \parallel \overline{TU}$

Example 1

In parallelogram ABCD, $\overline{AB} \parallel \overline{CD}$ and $\overline{AD} \parallel \overline{BC}$. If $\overline{AB} = 23.4$ and $\overline{BC} = 19.5$, find the perimeter of the parallelogram.

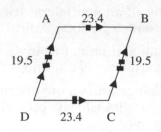

Analysis

Let P = the perimeter of the parallelogram.

In parallelogram ABCD, the opposite sides are congruent, so just find all the sides and add them up.

Work

$$P = \overline{AB} + \overline{BC} + \overline{CD} + \overline{AD}$$

$$P = 23.4 + 19.5 + 23.4 + 19.5 = 85.8$$

Answer

85.8

Exercises

1a. In parallelogram RSTU, $\overline{RS} = 14.8$ and $\overline{ST} = 13.6$. Find the perimeter.

1b. In parallelogram GHIJ, $\overline{HI} = 7.9$ and $\overline{IJ} = 12.4$. Find the perimeter.

Example 2

In parallelogram RSTU, the perimeter equals 56.4 and side $\overline{RS} = 12.3$. Find side $\overline{ST}$.

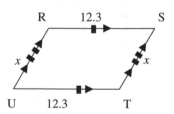

Analysis

Let P = the perimeter of the parallelogram.

$\overline{RS} \cong \overline{TU} = 12.3$.

Let $\overline{ST} \cong \overline{RU} = x$.

Work

$$P = \overline{RS} + \overline{ST} + \overline{TU} + \overline{RU}$$

$P = 56.4, \overline{RS} \cong \overline{TU} = 12.3,$

$\quad \overline{ST} \cong \overline{RU} = x:$ $\qquad 56.4 = 12.3 + x + 12.3 + x$

$\qquad\qquad\qquad\qquad\qquad 56.4 = 24.6 + 2x$

Subtract 24.6: $\qquad\qquad 31.8 = 2x$

Divide by 2: $\qquad\qquad 15.9 = x$

Symmetric Property: $\qquad x = 15.9 \left(\overline{ST} \right)$

Answer

$\quad \overline{ST} = 15.9$

2a. The perimeter of parallelogram KMNO is 45 and $\overline{MN} = 9.4$. Find side $\overline{KM}$.

2b. Side $\overline{BC}$ of parallelogram ABCD is 23. If the perimeter of the parallelogram is 72, find side $\overline{CD}$.

Exercises

Example 3

Side $\overline{AB}$ of parallelogram ABCD is 2 less than four times side $\overline{BC}$. If the perimeter is 56, find side $\overline{CD}$.

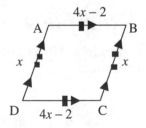

Analysis

Draw parallelogram ABCD and label the sides in alphabetical order.

Let $\overline{BC} \cong \overline{AD} = x$.

Let $\overline{AB} \cong \overline{CD} = 4x - 2$.

Let $P =$ the perimeter of the parallelogram.

Add up all the sides and set the total equal to the perimeter, 56.

Work

$$P = x + x + (4x - 2) + (4x - 2)$$

$P = 56$: $56 = 10x - 4$

Add 4: $60 = 10x$

Divide by 10: $6 = x$

Symmetric Property: $x = 6$

$$\overline{CD} = 4x - 2 = 4(6) - 2$$
$$= 24 - 2 = 22$$

Check

$$P = x + x + (4x - 2) + (4x - 2)$$

$P = 56, x = 6$: $56 = 6 + 6 + (24 - 2) + (24 - 2)$

$56 = 12 + 22 + 22$

$56 = 56$ ✓

Answer

$\overline{CD} = 22$

Exercises

3a. The perimeter of parallelogram EFGH is 34. If side $\overline{EF}$ is 2 more than twice side $\overline{FG}$, find side $\overline{EH}$.

3b. Side $\overline{BC}$ of parallelogram ABCD is 1 less than twice side $\overline{CD}$. If the perimeter is 40, find side $\overline{AD}$.

Example 4

Side $\overline{AB}$ of rhombus ABCD is 23.5. Find the perimeter of the rhombus.

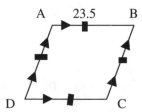

Analysis

Let P = the perimeter of the rhombus.

All the sides of the rhombus are equal, so just add them up to find the perimeter.

Work

$$P = \overline{AB} + \overline{BC} + \overline{CD} + \overline{DA}$$

$\overline{AB} \cong \overline{BC} \cong \overline{CD}$

$\cong \overline{DA} = 23.5$: $\quad P = 23.5 + 23.5 + 23.5 + 23.5$

$$P = 94$$

Answer

94

4a.   Side $\overline{FG}$ of rhombus EFGH is 26.4. Find the perimeter of the figure.

4b.   Side $\overline{RU}$ of rhombus RSTU is 18.3. What is the perimeter of the rhombus?

Exercises

Example 5

In rhombus ABCD, the perimeter is 22.8. Find one side.

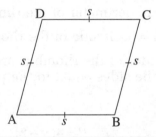

Analysis

Let P = the perimeter of the rhombus.

Let s = each side of the rhombus.

Since all the sides of the rhombus are equal, just divide the perimeter by 4 to find one side.

Work

$$P = 4s$$

$P = 22.8$: $\qquad\qquad 22.8 = 4s$

Divide by 4: $\qquad\qquad 5.7 = s$

Symmetric Property: $\qquad s = 5.7$

Answer

5.7

5a. The perimeter of rhombus GHIJ is 74.8. Find side IJ.

5b. If the perimeter of rhombus KLMN is 29.6, find side MN.

Example 6

If one side of a rhombus is represented by $x - 2$ and the perimeter is 36, find x.

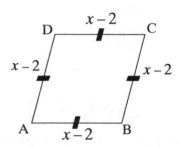

Analysis

Let P = the perimeter of the rhombus.

Let $x - 2$ = each side of the rhombus.

All the sides of the rhombus are equal. Set the total of all the sides equal to the perimeter, 36.

Work

$$P = 4(x - 2)$$

$P = 36$: $\qquad 36 = 4(x - 2)$

$\qquad\qquad\qquad 36 = 4x - 8$

Add 8: $\qquad\quad 44 = 4x$

Divide by 4: $\quad 11 = x$

Symmetric Property: $\quad x = 11$

Answer

11

6a. Side $\overline{BC}$ of rhombus ABCD is represented by $2x + 2$, and the perimeter of the figure is 32. Find x.

6b. One side of a rhombus RSTU is represented by $3x - 1$, and the perimeter is 44. Find x.

Example 7

In isosceles trapezoid RSTU, $\overline{RS} \parallel \overline{TU}$ and $\overline{ST} = \overline{RU}$. If the perimeter is 43, $\overline{ST} = 8$, and $\overline{RS} = 12$, find $\overline{TU}$.

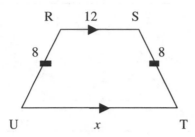

Analysis

Let P = the perimeter of the rhombus.

Let $x = \overline{TU}$.

$\overline{ST} \cong \overline{RU} = 8$ since the nonparallel sides in an isosceles trapezoid are equal.

Work

$$P = \overline{RS} + \overline{ST} + \overline{TU} + \overline{RU}$$

$P = 43$, $\overline{ST} \cong \overline{RU} = 8$,

$\overline{RS} = 12$, $\overline{TU} = x$: $43 = 12 + 8 + x + 8$

$43 = 28 + x$

Subtract 28: $15 = x$

Symmetric Property: $x = 15 \,(\text{side } \overline{TU})$

Answer

15

Exercises

7a.   In isosceles trapezoid ABCD, $\overline{AB} \parallel \overline{CD}$ and $\overline{AD} \cong \overline{BC}$. If the perimeter is 46, side $\overline{BC}$ is 9, and side $\overline{CD}$ is 15, find side $\overline{AB}$.

7b.   In isosceles trapezoid EFGH, $\overline{EF} \parallel \overline{GH}$ and $\overline{FG} \cong \overline{EH}$. If side $\overline{FG}$ is 8, side $\overline{EF}$ is 12, and the perimeter is 50, find side $\overline{GH}$.

Example 8

The perimeter of an isosceles trapezoid is 52. If each of the equal sides is 5 more than the smaller base and the larger base is 2 more than twice the smaller base, find each side of the figure.

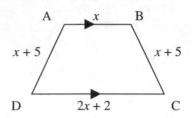

Analysis

Let P = the perimeter of the isosceles trapezoid.

Let x = the smaller base.

Let $2x + 2$ = the larger base.

Let $x + 5$ = each of the two equal sides.

Work

$$P = \overline{AB} + \overline{BC} + \overline{CD} + \overline{AD}$$

$P = 52, \overline{BC} \cong \overline{AD} = x + 5,$

$\overline{AB} = x, \overline{CD} = 2x + 2:$

$$52 = (x) + (x + 5) + (2x + 2)$$
$$+ (x + 5)$$
$$52 = 5x + 12$$

Subtract 12: $40 = 5x$

Divide by 5: $8 = x$

Symmetric Property: $x = 8 \left(\overline{AB} \right)$

$$x + 5 = 13 \left(\overline{BC} \text{ and } \overline{AD} \right)$$
$$2x + 2 = 18 \left(\overline{CD} \right)$$

Check

$$P = \overline{AB} + \overline{BC} + \overline{CD} + \overline{AD}$$

$x = 8, x + 5 = 13,$

$2x + 2 = 18:$

$$52 = (x) + (x + 5) + (2x + 2) + (x + 5)$$
$$52 = 8 + 13 + 18 + 13$$
$$52 = 52 \checkmark$$

Answer

$\overline{AB} = 8, \overline{BC} = 13, \overline{CD} = 18, \overline{AD} = 13$

8a. The perimeter of isosceles trapezoid RSTU is 35. $\overline{RS} \parallel \overline{TU}$ and $\overline{ST} = \overline{RU}$. If side $\overline{TU}$ is 1 less than twice side $\overline{RS}$ and side $\overline{ST}$ is 3 more than side $\overline{RS}$, find side $\overline{RU}$.

8b. In isosceles trapezoid LMNO, $\overline{LM} \parallel \overline{NO}$ and $\overline{MN} \cong \overline{LO}$. Side $\overline{MN}$ is 4 more than side $\overline{LM}$, and side $\overline{NO}$ is 5 less than side $\overline{LM}$. If the perimeter is 43, find side $\overline{LM}$.

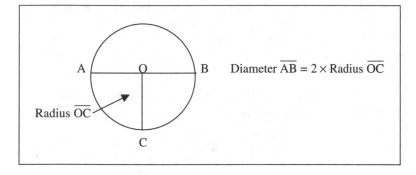

Diameter $\overline{AB} = 2 \times$ Radius $\overline{OC}$

Radius $\overline{OC}$

It's as easy to determine the measure of the radius of a circle as it is to find its diameter. However, it's a bit harder to determine the circumference of the circle. Primitive societies simply laid out a string around the circumference of a circle, cut it, and then placed the string along a straight edge. The circumference, then, could be easily measured along the straight edge.

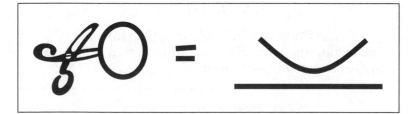

More advanced civilizations realized that there was a relationship between the circumference and the diameter. In the Old Testament, the estimate was that the circumference was equal to three times the diameter. The Babylonians estimated that the circumference was $3\frac{1}{8}$ times the diameter while the ancient Egyptians calculated the multiple as $\frac{256}{81}$.

Archimedes (c. 287–212 B.C.E.) was probably the greatest mathematician of the Hellenistic period (4th–1st century B.C.E.). He was the court mathematician for Hiero, the king of Syracuse.

Archimedes approximated the circumference of a circle by using inscribed and circumscribed regular polygons. Eventually, he used a 96-sided regular polygon to estimate the circumference of a circle. He called the ratio of the circumference to the diameter π, and he came closest to our own measure for π:

$$\pi = \frac{C}{d} = \frac{22}{7} = 3\frac{1}{7} \approx 3.14$$

Eventually, mathematicians proved that π is an **irrational number**—it cannot be expressed as a ratio of two whole numbers. Rather, its decimal expansion goes on forever without repeating itself. As of the year 2001, computers estimated the value of π to over 1.1 billion places.

Let's return to the Greek formula and find the value of C in terms of d:

$$\pi = \frac{C}{d}$$

Multiply by d: $d\left(\pi = \frac{C}{d}\right)$

$$\pi d = C$$

Symmetric Property: $C = \pi d$

Example 1

The Coliseum in Rome is approximately 560 feet in diameter. Find its circumference. Round to the nearest tenth of a foot.

Analysis

Let C = the circumference of the circle.

Let $\pi = 3.14$.

Let d = the diameter of the circle.

Use the formula $C = \pi d$ to find the circumference.

Work

$$C = \pi d$$

$\pi = 3.14,\ d = 560$: $C = 3.14(560) = 1758.40 \approx 1758.4$

Answer

    1,758.4 feet

1a. A circular garden has a diameter of 34 feet. Jason wants to enclose the garden with a fence. To the closest foot, find the length of the fence necessary to enclose the garden. Let $\pi = 3.14$.

1b. The diameter of an igloo is 14 feet. Find its circumference. Round to the nearest foot. Use $\pi = 3.14$.

Exercises

Example 2

    If the radius of a circle is 8.6, find its circumference. Round to the nearest tenth.

Analysis

    Let C = the circumference of the circle.

    Let $\pi = 3.14$.

    Let d = the diameter of the circle.

    Use the formula $C = \pi d$.

Work

$$C = \pi d$$

$r = 8.6,\ d = 2r$: $C = \pi(2r) = \pi(2 \times 8.6) = 17.2\pi$

$\pi = 3.14$: $C = 17.2(3.14) = 54.008 \approx 54.0$

Answer

    54.0

2a. The radius of a circle is 5 in. To the nearest inch, find the circumference. Use $\pi = 3.14$.

2b. Find the circumference of a circle if its radius is 6.3 feet. Round to the nearest foot. Use $\pi = 3.14$.

Exercises

Example 3

If the circumference of a circular racetrack is $2\frac{5}{14}$ miles, find its diameter.

Analysis

Let C = the circumference of the circle.

Let $\pi = \frac{22}{7}$.

Let d = the diameter of the circle.

Use the formula $C = \pi d$ and substitute $2\frac{5}{14}$ for C.

Work

$$C = \pi d$$

$$C = 2\frac{5}{14}, \pi = \frac{22}{7} \qquad 2\frac{5}{14} = \frac{22}{7}d$$

$$\frac{33}{14} = \frac{22}{7}d$$

Multiply by 14: $\qquad 14\left(\frac{33}{14} = \frac{22}{7}d\right)$

$$33 = 44d$$

Divide by 44: $\qquad \frac{33}{44} = d$

Symmetric Property: $\qquad d = \frac{33}{44} = \frac{3}{4}$

Answer

$\frac{3}{4}$ miles

Exercises

3a. If the circumference of a circle is $15\frac{5}{7}$, find its diameter. Use $\pi = \frac{22}{7}$.

3b. What is the radius of a circle whose circumference is 30.144. Let $\pi = 3.14$.

Example 4

If a car is driven 2,512 feet and its tires turn 200 revolutions in that period, what is the circumference of each tire?

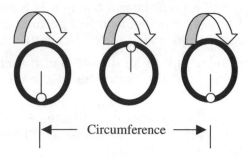

|← Circumference →|

Analysis

During 200 revolutions, the tires cover 2,512 feet.

Let C = the circumference of the tire.

Work

To find the circumference, we'll just divide the distance by the number of revolutions:

$$C = \frac{2,512}{200} = 12.56$$

Answer

12.56 feet

Exercises

4a.   Tanequa drives a bike 11,605.44 feet. If the wheels turn 1,232 times, find the circumference of the wheel.

4b.   The wheels on a motorcycle have a circumference of 8.76 ft. If Helmut drives the motorcycle 17,896.68 feet, how many rotations did the wheels make?

Test

$C = 2\pi r$

1–16. @ 5% each

1.   The sides of a hexagon-shaped area are 55.4 feet, 69.7 feet, 48.5 feet, 87 feet, 49.2 feet, and 79.9 feet. Find the cost of constructing a fence around the perimeter if the fence costs $2.58 per foot. Round your answer to the nearest cent.

2.   The sides of a pentagon are represented by x, $x + 2$, $x + 5$, $2x + 2$ and $3x - 1$. If the perimeter is 56, find each side.

3.   Each side of a regular hexagon is $x + 7$. If the perimeter is 66, find x.

4. Each side of an equilateral triangle is 5 more than each side of a regular octagon. If the perimeters of both figures are equal, find a side of the octagon.

5. A rectangular plot of land is 45 feet 8 inches by 72 feet 5 inches. Find the perimeter of the plot, in feet and inches.

6. The length of a rectangular ballroom is 5 feet less than twice the width. If the perimeter is 80 feet, find the dimensions of the ballroom.

7. The base of an isosceles triangle is 36 and its perimeter is 102. Find one of the equal sides.

8. The perimeter of an equilateral triangle is 164.1. Find one side.

9. One side of an equilateral triangle is 7.93. Find the perimeter.

10. The perimeter of an equilateral triangle is 57. If each side is represented by $3x + 1$, find x.

11. The perimeter of parallelogram ABCD is 34. If side $\overline{AB}$ is 2 more than twice side $\overline{BC}$, find side $\overline{AD}$.

12. Side $\overline{RS}$ of rhombus RSTU is represented by $4x + 1$, and the perimeter of the figure is 68. Find x.

13. A circular botanical garden has a diameter of 92 feet. Howser wants to enclose the garden with a fence. How many feet of fence are necessary? Let $\pi = 3.14$ and round the answer to the nearest foot.

14. Find the circumference of a circle if its radius is 7.4. Use $\pi = 3.14$ and round the answer to the nearest foot.

15. What is the radius of a circle whose circumference is 150.72? Let $\pi = 3.14$.

16. The wheels on a car have a circumference of 10.99 feet. If Max drives the car 26,376 feet, how many revolutions did the wheels make?

17–18. @ 10% each

17. If the perimeter of an isosceles triangle is 67 and each of the equal sides is 13 less than the base, find each side of the triangle.

18. The perimeter of isosceles trapezoid KLMN is 45. $\overline{KL} \parallel \overline{MN}$ and $\overline{KN} = \overline{LM}$. If side $\overline{LM}$ is 4 more than side $\overline{KL}$ and side $\overline{MN}$ is 3 less than twice side $\overline{KL}$, find side $\overline{MN}$.

4

Areas

Area is the measure of an enclosed two-dimensional region. Area is expressed as the number of square units within the region (square feet, square inches, square meters, etc.).

A **rectangle** is a quadrilateral with four right angles. Let's determine the number of units within the rectangle—the area. One way of doing this is to just count the number of boxes inside the rectangle—in this case (see page 88), 32.

(see page 88)

A more efficient way of determining the area is to multiply length (*l*) times width (*w*).

$$A = lw$$
$$A = 8(4) = 32$$

4.1 Rectangles

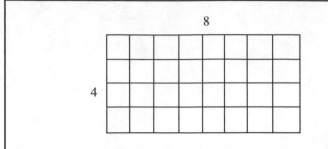

8

4

Example 1

A parking lot has to be paved at a cost of $1.48 per *square foot*. If the lot is 120 feet × 90 feet, how much does it cost to pave the lot?

Analysis

Let A = area of rectangle.

Let l = length of rectangle.

Let w = width of rectangle.

Find the area of the lot and then multiply the number of square feet by the cost of paving each square foot, $1.48.

Work

$$A = lw$$

$l = 120, w = 90$: $\qquad A = (120)(90) = 10{,}800$

Multiply the area by
$1.48: $\qquad 10{,}800 \times \$1.48 = \$15{,}984$

Answer

$15,984

Exercises

1–6. $A_{\text{rectangle}} = bh$

1a. Janet is going to wallpaper her wall. If the wall measures 15 feet by 11 feet and wallpaper costs $0.64 per square foot, how much will the wallpaper cost?

1b. A floor measures 23 feet by 16 feet. If Jenna wants to have it varnished and the cost is $1.02 per square foot, how much would it cost to varnish the entire floor?

Example 2

A bed of earth is to be planted around a pool. If the bed of earth is 45 feet long and 34 feet wide and the pool is 31 feet long and 26 feet wide, how many square feet of earth do we need?

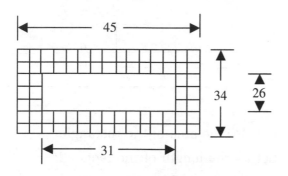

Analysis

Let's find the number of square feet in the cross-hatched area. To do that, we'll first find the outside area and then subtract the inner pool area.

Work

$$\text{Area} = \text{Length} \times \text{Width}$$
$$A = lw$$

The outside area:	$1,530 = 45 \times 34$
− The inner area:	$-806 = -31 \times 26$
The cross-hatched area:	724

Answer

724 square feet

2a.   The area around a pool is to be tiled. If the pool measures 34 feet by 28 feet and the tiled area around the pool measures 39 feet by 33 feet, how many square feet of tile are needed?

2b.   Shrubbery is supposed to be planted around a new building. If the building's base is in the shape of a rectangle 75 feet by 69 feet and the shrubbery is to extend 81 feet by 75 feet around the building, how many square feet of shrubbery are needed?

Exercises

Example 3.

Workers are enclosing a playground with a fence. If the area of the playground is 4,723.92 square feet and the width is 64.8 feet, what is the length of the playground?

$$l$$

| $A = 4{,}723.92$ | $w = 64.8$ |

Analysis

Let A = the area of the rectangle.

Let l = the length of the rectangle.

Let w = the width of the rectangle.

In the formula $A = lw$, just substitute 4,723.92 for A and 64.8 for w.

Work

$$A = lw$$

$$4{,}723.92 = l(64.8)$$

Divide by 64.8: $72.9 = l$

Answer

72.9 feet

Exercises

3a. The floor of a warehouse is 6,828.4 square feet. If the width is 79.4 feet, find its length.

3b. A desktop measures 5,032 square inches. If its length is 74 inches, what is its width?

Example 4

Zanzibar Realty owns a rectangular commercial loft. The length is 10 feet less than 3 times the width and the width is 50 feet. If Zanzibar gets a monthly income of $1.25 per square foot, how much is the total monthly rental for this commercial loft?

Analysis

Let A = the area of the rectangle.

Let w = the width of the rectangle.

Let $3w - 10$ = the length of the rectangle.

Find the number of square feet in the loft and then multiply by $1.25 the income per square foot.

Work

The length is 10 feet less than 3 times the width:

$$length = 3w - 10$$

$w = 50$: $\quad length = 3(50) - 10 = 140$

$$Area = lw$$

$$A = lw$$

$l = 140$, $w = 50$: $\quad A = (140)(50) = 7{,}000$ square feet

Multiply by $1.25 per sq ft: $\quad \$1.25 \times 7{,}000 = \$8{,}750$

Answer

$8,750

4a.   The length of a living room is 4 feet more than its width. The width is 13 feet. If the cost of shellacking the floor is $1.10 per square foot, how much does it cost to shellac the living room floor?

4b.   Julio wants to sod his backyard. The length of the yard is 12 feet less than three times the width. If the width is 28 feet and the sod costs $0.56 per square foot, how much would it cost to sod his backyard?

Exercises

Example 5

The Karistex Corporation is selling rugs at $15.18 per square *yard*. Elka wants to purchase a rug for her living room that measures 18 *feet* by 15 *feet*. How much will such a rug cost?

Analysis

Let A = the area of the rectangle.

Let l = the length of the rectangle.

Let w = the width of the rectangle.

First, change the dimensions of the living room to yards. Then, find the area in square yards and, finally, multiply the number of square yards by $15.28, the cost per square yard. (Three feet equal one yard.)

Work

$$\frac{18}{3} \text{ feet} = 6 \text{ yards} \qquad \frac{15}{3} \text{ feet} = 5 \text{ yards}$$

In yards, the dimensions of the room are 6 yards by 5 yards.

$$A = lw$$

$l = 6$, $w = 5$: $\qquad A = 6 \times 5 = 30$ square yards

Multiply the area by $15.18, the per square yard cost of the rug: $\qquad 30 \times \$15.18 = \$455.40$

Answer

$455.40

Exercises

5a. Wallpaper costs $1.89 per square *yard*. If a wall measures 18 *feet* by 21 *feet*, how much will it cost to wallpaper the entire wall?

5b. It costs $4.73 per square *yard* to put up a new roof. If Bob's roof measures 33 *feet* by 51 *feet*, how much would it cost to put up a new roof?

Example 6

The area of a rectangle is 224. If its length is represented by $5x - 2$ and its width is 8, find x.

Analysis

Let A = the area of the rectangle.

Let $5x - 2$ = the length of the rectangle.

Work

$$A = lw$$

$A = 224$, $l = 5x - 2$, $w = 8$: $\qquad 224 = (5x - 2)(8)$

$$224 = 40x - 16$$

Add 16: $240 = 40x$

Divide by 40: $6 = x$

Symmetric Property: $x = 6$

Answer

  $x = 6$

6a.   The width of a room is 14 feet and the length is represented by $2x - 10$ feet. If its area is 252 square feet, find x.

Exercises

6b.   The width of a rectangular lot is 47 and the length is represented by $3x - 4$. If the area is 2,632, find x.

A **parallelogram** is a quadrilateral with both pairs of opposite sides parallel. Given the formula for the area of a rectangle, let's figure out the formula for the area of a parallelogram. We'll start out with parallelogram ABCD.

4.2
Parallelograms

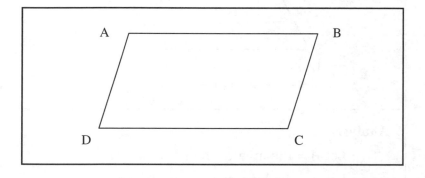

Let's draw $\overline{AF}$ perpendicular to base $\overline{CD}$ and $\overline{BE}$ perpendicular to base $\overline{CD}$ extended to E. We've created two new triangles, △I and △II, and they're both equal in area.

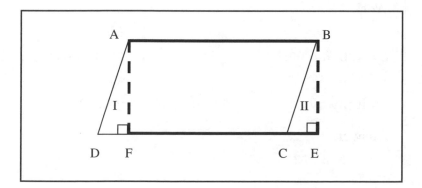

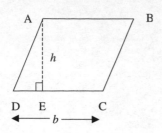

We have ABCF + △I ≅ ABCF + △II or parallelogram ABCD ≅ rectangle ABEF. Since parallelogram ABCD and rectangle ABEF are congruent, their areas are equal, so that we can now use the formula $A = lw$ or, in the case of a parallelogram, $A = bh$, where A = area, b = base, and h = height. In the case of a parallelogram, the height is determined by drawing a perpendicular from one vertex to the opposite base.

Example 1

Mr. Thompson wants to purchase a piece of property in the shape of a parallelogram. The base of the property is 120 feet and its height is 95 feet. If the property costs $2.48 per square foot, what is its full cost?

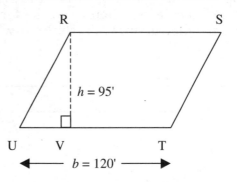

Analysis

Let A = the area of the parallelogram.

Let b = the base of the parallelogram.

Let h = the height of the parallelogram.

To determine the area, use the formula $A = bh$. After you find the area, multiply by $2.48, the cost per square foot.

Work

$$A = bh$$

$b = 120, h = 95$:

$$A = (120)(95)$$

$$A = 11,400$$

Multiply by $2.48: $(11,400)(2.48) = 28,272$

Answer

$28,272

1–3.   Area of a parallelogram = base × height

1a.   Heinrich wants to purchase a piece of silver in the shape of a parallelogram to make a piece of jewelry. The piece has a base of 4.5 inches and a height of 3.8 inches. Find its cost if it sells for $2.25 per square inch. Round the answer to the nearest cent.

1b.   A city lot in the shape of a parallelogram has a base of 104 feet and a height of 73 feet. If the property sells for $4.96 per square foot, what is the total cost?

Example 2

ABCD is a picture of a parcel of land in the shape of a parallelogram. The base, $\overline{CD}$, is 130 feet, and the area of the parallelogram is 5,980 square feet. Find the height, h.

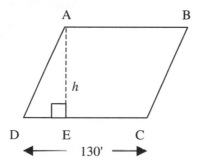

Analysis

Let A = the area of the parallelogram.

Let b = the base of the parallelogram.

Let h = the height of the parallelogram.

Use the formula $A = bh$. Substitute 130 for the base, b, and 5,980 for the area.

Work

Area = Base × Height: $A = bh$

$A = 5,980$, $b = 130$: $5,980 = (130)h$

Divide by 130: $46 = h$

Symmetric Property: $h = 46$

Answer

46 ft

2a.   Parallelogram ABCD is 488.3 square feet in area. If its height is 19, find its base.

2b.   The area of parallelogram ABCD is 151.68. If its base is 15.8, find its height.

Example 3

A parallelogram is 9 units high. If $x + 2$ represents the base and the area is 108 square units, find x.

Analysis

Let A = the area of the parallelogram.

Let $x + 2$ = the base of the parallelogram.

Let 9 = the height of the parallelogram.

Work

Area = Base × Height

$A = 108, b = x + 2, h = 9$: $108 = (x + 2)9$

$108 = 9x + 18$

Subtract 18: $90 = 9x$

Divide by 9: $10 = x$

Answer

10

3a.   The area of parallelogram ABCD is 665. If the base is represented by $2x - 1$ and the height is 19, find x.

3b.   The base of a parallelogram is represented by $2x + 1$. If the area is 171 and the height is 9, find x.

4.3 Squares

A **square** is a rectangle whose sides are of equal length. Let's say that we have a square, 6 units long by 6 units wide. What is its area—the number of units within the figure?

One way to determine the number of units is simply to count them. This method is time consuming but accurate. If we take up this suggestion, we count 36 units.

It's much more convenient to use a formula—especially when we have more complicated figures. So, if we multiply the length by the width, 6 × 6, we also arrive at 36 units.

If A represents the area of a square and s represents each side, then the area equals side × side:

$$A = s \cdot s = s^2$$

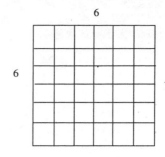

Example 1

Find the area of a square whose side is $4\frac{3}{8}''$.

Analysis

Use the formula $A = s^2$.

Work

$$A = s^2 = \left(4\frac{3}{8}\right)^2 = \frac{35}{8} \times \frac{35}{8} = \frac{1{,}225}{64} = 19\frac{9}{64}$$

Answer

$19\frac{9}{64}$ square inches

$A_{\text{square}} = s^2$

1a. Find the area of a square whose side is $3\frac{4}{5}$.

1b. Find the area of a square whose side is 7.4.

Exercises

Example 2

Find the cost of carpeting a square living room 16.5 feet on each side if the carpet costs $2.39 per square foot. Let's round the answer to the nearest cent.

Analysis

Let A = the area of the square.

Let s = a side of the square.

First, we'll use the formula $A = s^2$ to find the area. After we have the area in square feet, we'll multiply by $2.39, the cost per square foot of carpet.

Work

$$A = s^2 = (16.5)^2 = 16.5 \times 16.5 = 272.25 \text{ square feet}$$

Find the cost of the carpet:
$$\$2.39 \times 272.25 = \$650.6775 \approx \$650.68$$

Answer

$650.68

Exercises

2a.   The side of a square field measures 132 feet. Jeremy wants to cover the field with new sod. If sod costs $0.27 per square foot, what is the total cost for the sod?

2b.   A storefront window is broken and must be replaced. The window is a square and measures 14 feet on a side. If replacement glass costs $0.89 per square foot, what is the total cost for a new window?

Example 3

Find one side of a square if its area = 204.49 square feet.

Analysis

Use the formula $A = s^2$, where A = area and s = side. Then, find the square root of A.

Work

$$A = s^2$$

$A = 204.49$: $\qquad\qquad 204.49 = s^2$

Take the square root: $\qquad \sqrt{204.49} = s$

$$14.3 = s$$

Symmetric Property: $\qquad\qquad s = 14.3$

Answer

14.3 feet

Exercises

3a.   Find a side of a square if its area is 823.69.

3b.   How large is a side of a square whose area is 7.84?

Example 4

If each side of a square is 9 and the area is represented by $5x + 46$, find x.

Analysis

Let $5x + 46$ = the area of the square.

Let 9 = a side of the square.

Work

$$A = s^2$$

$A = 5x + 46$, $s = 9$: 　$5x + 46 = 9^2$

$$5x + 46 = 81$$

Subtract 46: 　　　　$5x = 35$

Divide by 5: 　　　　$x = 7$

Answer

$x = 7$

Exercises

4a. A side of a square is 11. If the area is represented by $8x + 25$, find x.

4b. The area of a square is represented by $20x - 56$. If each side is 18, find x.

A **triangle** is a three-sided polygon. Let's derive the formula for the area of a triangle from something we already know, the area of a parallelogram.

4.4 Triangles

Let's begin with triangle ABC. We'll start by constructing two triangles, $\triangle A'B'C'$ and $\triangle A'C'D'$, identical to our original triangle, $\triangle ABC$. Let's arrange it so that the two new triangles share the same adjoining side, $\overline{A'C'}$. A new figure is formed, parallelogram $A'B'C'D'$, with an area double the original triangle, $\triangle ABC$.

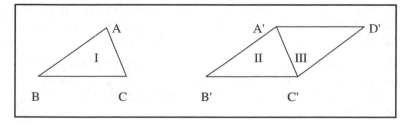

We already know that the area of a parallelogram is equal to the base times the height, so now we can see that the area of $\triangle$I equals half of the area of parallelogram A′B′C′D′. We can now conclude that Area $\triangle = \frac{1}{2}b \cdot h$, where b = base and h = height.

Example 1

Mr. James wants to cut out a triangular piece from a sheet of gold for a piece of jewelry he wants to design. Gold from the sheet costs $11.92 per square inch. If Mr. James wants to design a triangular piece of jewelry with a base of 1.5 inches and a height of 2.4 inches, how much would the triangular piece cost? Round the answer to the nearest cent.

Analysis

Let A = the area of the triangle.

Let b = the base of the triangle.

Let h = the height of the triangle.

Use the formula $A = \frac{1}{2}bh$. After you have determined the area, multiply by $11.92, the cost per square inch.

Work

$$A = bh$$

$b = 1.5, h = 2.4$: $A = (0.5)(1.5)(2.4) = 1.8$

Multiply area
by the cost per
square inch: $(1.8)(11.92)$ $= 21.456 \approx 21.46$

Answer

$21.46

Exercises

$A_{\text{triangle}} = \frac{1}{2}bh$

1a. Hillary wants to place marble tiles over a triangular area. The base is 27 feet, and the height is 18 feet. If the cost of tiles is $2.27 per square foot, how much will it cost to cover the area in tiles?

1b. Linda has a triangular piece of property with a base of 84 feet and a height of 73 feet. If the property sells for $3.21 per square foot, how much is the property worth?

Example 2

If the area of a triangle is 117 square inches and its height is 9 inches, find the base.

Analysis

Let A = the area of the triangle.

Let b = the base of the triangle.

Let h = the height of the triangle.

Use the area formula $A = \frac{1}{2}bh$ and substitute the given information.

Work

$$A = \tfrac{1}{2}bh$$

$A = 117,\ h = 9$: $\qquad 117 = (0.5)(b)(9)$

$\qquad\qquad\qquad\qquad 117 = 4.5b$

Divide by 4.5: $\qquad\quad 26 = b$

Symmetric Property: $\qquad b = 26$

Answer

26 inches

2a.   If the area of a triangle is 189 and its base is 27, find its height.

2b.   The height of a triangle is 32 and its area is 2,048. Find its base.

Exercises

Example 3

The base of a triangle is represented by $3x - 1$. If the area is 140 and the height is 20, find x.

Analysis

Let A = the area of the triangle.

Let $3x - 1$ = the base of the triangle.

Let 20 = the height of the triangle.

Use the area formula and then substitute the given information.

Work

$$A = \tfrac{1}{2}bh$$

$A = 140,\ b = 20,\ h = 3x - 1:\qquad 140 = \tfrac{1}{2}(20)(3x - 1)$

$$140 = 10(3x - 1)$$

$$140 = 30x - 10$$

Add 10: $150 = 30x$

Divide by 30: $5 = x$

Symmetric Property: $x = 5$

Answer

    5

Exercises

3a.   The height of a triangle is represented by $x + 7$. If the area is 120 and the base is 12, find x.

3b.   The area of a triangle is 242. If the base is 11 and the height is represented by $2x + 2$, find x.

Example 4

Ms. Combs owns a triangular piece of property with a base of 290 feet and a height of 167 feet. Ms. Holloway owns a piece of property in the shape of a parallelogram with a base of 134 feet and a height of 140 feet. Which piece of property is larger, and by how much?

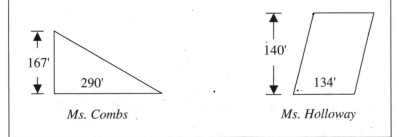

Ms. Combs   Ms. Holloway

Analysis

For Ms. Combs' triangular property, use the formula $A = \tfrac{1}{2}bh$, where A = area, b = base, and h = height. For Ms. Holloway's property in the shape of a parallelogram, use the formula $A = bh$, where A = area, b = base, and h = height.

Work

Ms. Combs: $\text{Area}_{\text{Triangle}} = \frac{1}{2}\text{Base} \times \text{Height}$

$$A = \tfrac{1}{2}bh$$

$b = 290$, $h = 167$: $A = (0.5)(290)(167) = 24{,}215$

Ms. Holloway:

$$\text{Area}_{\text{Parallelogram}} = \text{Base} \times \text{Height}$$

$$A = bh$$

$b = 134$, $h = 140$: $A = (134)(140) = 18{,}760$

Difference: $24{,}215 - 18{,}760 = 5{,}455$

Answer

Ms. Combs' property is larger by 5,455 square feet.

Exercises

4a. The base of a rectangle is 14, and its height is 9. The base of a triangle is 18, and its height is 15. Which figure is larger in area, and by how much?

4b. A triangle has a base of 18 and a height of 12. A square has a side of 14. Which figure is larger? By how much?

Example 5

Ms. Ramos wants to design a metal sign in the shape of a triangle. The triangle, with a base of 14.2″ and a height of 17.9″ is cut from a rectangular piece of metal in the dimensions of 14.2″ by 17.9″.

a. In area, how much material is wasted?

b. If the metal costs $0.09 per square inch, what is the cost of the wasted material? Round to the nearest cent.

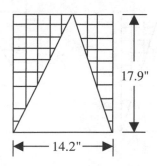

17.9"

14.2"

Analysis

The cross-hatched area represents the wasted material. To find the cross-hatched area, we'll subtract the area of the triangle from the area of the rectangle. Then, we'll multiply the difference by $0.09, the cost per square inch.

Work

Area of rectangle = Base × Height:
$$A_R = bh = 14.2 \times 17.9 = 254.18$$

– Area of triangle = $\frac{1}{2}$Base × Height:
$$- A_T = \tfrac{1}{2}bh = \tfrac{1}{2}14.2 \times 17.9 = 127.09$$

Cross-hatched area: $\qquad\qquad\qquad\qquad$ 127.09

Cost is $0.09 per square inch:
$$127.09 \times 0.09 = 11.4381$$

Answer

    a.   127.09 square inches

    b. $11.44

Exercises

5a. A triangle is cut out of a metal rectangle. The triangle has a base of 23″ and a height of 38″, and the rectangle has the same base and height. Find the wasted material. If each square inch of metal costs $0.34, what is the cost of the wasted material?

5b. A triangle whose base is 8″ and whose height is 19″ is cut from a rectangular piece of silver with the same base and height. How much silver is wasted? At $2.63 per square inch, what is the cost of the wasted material?

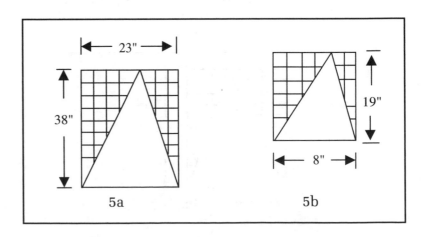

            5a   5b

In classical Greek time, the area of a circle was approximated by circumscribing regular polygons around the circle and simultaneously inscribing the same number of polygons inside the circle. It was relatively easy to determine the area of a polygon simply by dividing it into triangles and then adding up the areas of the triangles. As the number of sides of the polygon increases, the area of the polygon approaches the area of a circle.

Eventually, a more precise formula for the area of a circle was developed: $A = \pi r^2$, where A = the area, π = 3.14, and r = the radius.

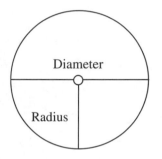

Example 1

An amusement park is planning a circular ride for children. If the radius of the circle is 19 feet, what is the area?

Analysis

Let A = the area of the circle.

Let π = 3.14.

Let r = the radius of the circle.

Use the formula $A = \pi r^2$.

Work

$$A = \pi r^2$$

$\pi = 3.14, r = 19$: $A = 3.14(19)^2$

$$A = 3.14 \times 361$$

$$A = 1,133.54$$

Answer

1,133.54 square feet

Area of a circle = πr^2

1a. The radius of the base of a circular building is 52 feet. Find the area of the base of the building. Let π = 3.14.

1b. If the radius of a circular tabletop is 3 feet, what is its area? Let π = 3.14.

Example 2

A belt fits around a disk that has a diameter of 18.6 inches. Find the area of the disk, to the nearest hundredth of an inch.

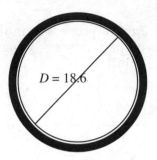

$D = 18.6$

Analysis

Let A = the area of the circle.

Let π = 3.14.

Let r = the radius of the circle.

Let d = the diameter.

Use the formula $A = \pi r^2$.

Work

Radius = $\frac{1}{2}$ diameter: $r = \frac{1}{2}d$

$d = 18.6$: $r = (0.5)(18.6) = 9.3$

Now, find the area
of a circle: $A = \pi r^2$

$\pi = 3.14$, $r = 9.3$: $A = 3.14(9.3)^2 = 3.14(86.49)$

$= 271.5786 \approx 271.58$

Answer

271.58 square inches

Exercises

2a.   The diameter of a circular pool is 24.6 feet. Let π = 3.14 and find the area of the pool, to the nearest square foot.

2b.   The diameter of a circle is 8.7. Let π = 3.14 and find the area, to the nearest tenth.

Example 3

A circular lawn has been covered with 5,024 square feet of sod.

    a. Find the radius of the lawn.

    b. Find the diameter of the lawn.

Analysis

Let A = the area of the circle.

Let π = 3.14.

Let r = the radius of the circle.

Use the formula $A = \pi r^2$. Substitute 5,024 for the area, A, to find the radius, r. Then double the radius to determine the diameter, D.

Work

$$A = \pi r^2$$

$A = 5{,}024, \pi = 3.14$:	$5{,}024 = 3.14 r^2$
Divide by 3.14:	$5{,}024 = 3.14 r^2$
	$1{,}600 = r^2$
Take the square root:	$40 = r$
Symmetric Property:	$r = 40$
Diameter = $2r$:	$D = 2 \times 40 = 80$

Answers

    a. Radius = 40 feet

    b. Diameter = 80 feet

3a. The area of a circle is 2,289.06. Let π = 3.14 and find the radius of the circle.

3b. If the area of a circle is 4,534.16 square feet, find its diameter. Let π = 3.14.

Exercises

Example 4

Manny wants to cut a circular table top from a square piece of marble. If the radius of the circle is 3 feet and the square piece of marble is 6 feet by 6 feet, how many square feet of marble are wasted? Use 3.14 for the value of π.

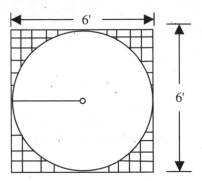

Analysis

The wasted area is the cross-hatched area, so we'll have to subtract the circular area from the square area to find the wasted area.

Work

Area of square = $s^2 = 6^2$ = 36.00

− Area of circle = πr^2 = 3.14 × 3^2 = 3.14 × 9 = 28.26

Wasted area = 7.74

Answer

7.74 square feet of marble are wasted.

Exercises

4a. A circular plate whose radius is 5 feet is to be cut from a square piece of metal, 10 feet on a side. How much material is wasted (the cross-hatched area)? Let π = 3.14.

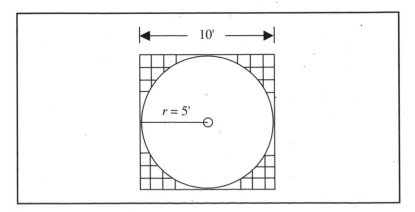

4b. A square, 4 feet on a side, is to be cut from a round piece of plastic whose radius is $\sqrt{8}$. Find the wasted area (the cross-hatched area). Let $\pi = 3.14$.

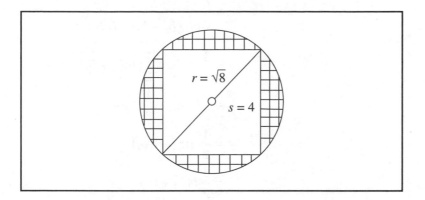

A **trapezoid** is a quadrilateral with two parallel sides and two nonparallel sides. The steel girders supporting bridges are often imbedded in concrete bases with sides shaped as trapezoids. It's been found that these trapezoidal shapes provide additional support for the girders.

4.6 Trapezoids

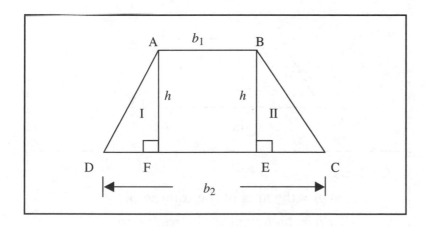

Let's develop a formula for the area of trapezoid ABCD. First, we'll draw perpendiculars $\overline{AF}$ and $\overline{BE}$ to base $\overline{CD}$ of the trapezoid. We've created two new triangles, $\triangle I$ and $\triangle II$. If we add up the areas of the two triangles plus the area of rectangle ABEF, we should have the formula for the area of the trapezoid.

$$\text{Area of } \triangle I = \frac{1}{2}(\overline{DF})(h)$$

$$\text{Area of } \triangle II = \frac{1}{2}(\overline{CE})(h)$$

Area of rectangle ABEF $= b_1 h$

Area of trapezoid ABCD $= \frac{1}{2}(\overline{DF})(h) + \frac{1}{2}(\overline{CE})(h) + b_1h$

$$= \frac{1}{2}(h)(\overline{DF} + \overline{CE}) + b_1h$$

$$= \frac{1}{2}(h)(b_2 - b_1) + b_1h$$

$$= \frac{1}{2}(h)b_2 - \frac{1}{2}(h)b_1 + b_1h$$

$$= \frac{1}{2}(h)b_2 + \frac{1}{2}(h)b_1$$

$$= \frac{1}{2}(h)(b_2 + b_1)$$

$$= \frac{1}{2}h(b_1 + b_2)$$

Example 1

Find the area of a trapezoid with an upper base of 9, a lower base of 15, and a height of 6.

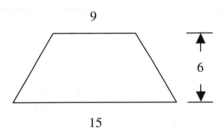

Analysis

Let A = the area of the trapezoid.

Let h = the height of the trapezoid.

Let b_1 = the upper base of the trapezoid.

Let b_2 = the lower base of the trapezoid.

Work

$$A = \tfrac{1}{2}h(b_1 + b_2)$$

$h = 6,\ b_1 = 9,\ b_2 = 15:$ $\quad A = \tfrac{1}{2}(6)(9 + 15)$

$$A = 3(24) = 72$$

Answer

72

Area of a trapezoid = $\frac{1}{2}$(Height)(Upper base + Lower base)

1a. Find the area of a trapezoid with a lower base of 24, an upper base of 16, and a height of 7.

1b. The two bases of a trapezoid are 15 and 37, and its height is 13. Find its area.

Example 2

The area of a trapezoid is 72. If the two bases are 7 and 17, find the height.

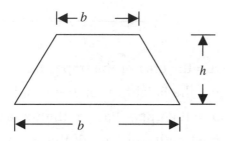

Analysis

Let A = the area of the trapezoid.

Let h = the height of the trapezoid.

Let b_1 = the upper base of the trapezoid.

Let b_2 = the lower base of the trapezoid.

Work

$$A = \tfrac{1}{2}h(b_1 + b_2)$$

$A = 72$, $b_1 = 7$, $b_2 = 17$: $\quad 72 = \tfrac{1}{2}h(7 + 17)$

Multiply by 2: $\quad 2[72 = \tfrac{1}{2}h(24)]$

$$144 = 24h$$

Divide by 24: $\quad 6 = h$

Symmetric Property: $\quad h = 6$

Answer

6

2a. The area of a trapezoid is 168. Its height is 12, and its upper base is 7. Find its lower base.

2b. If the lower base of a trapezoid is 31, its upper base is 19, and its area is 150, find its height.

Example 3

One base of a trapezoid is 17 and its height is 6. If the area is 87 and the shorter base is represented by $x - 3$, find x.

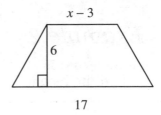

Analysis

Let A = the area of the trapezoid.

Let h = the height of the trapezoid.

Let b_1 = the upper base of the trapezoid.

Let b_2 = the lower base of the trapezoid.

Just use the formula $A = \frac{1}{2}h(b_1 + b_2)$ and substitute.

Work

$$A = \frac{1}{2}h(b_1 + b_2)$$

$A = 87$, $h = 6$, $b_1 = 17$, $b_2 = x - 3$:

$$87 = \frac{1}{2}(6)(17 + x - 3)$$
$$87 = 3(14 + x)$$
$$87 = 42 + 3x$$

Subtract 42: $\qquad 45 = 3x$

Divide by 3: $\qquad 15 = x$

Symmetric Property: $\qquad x = 15$

$$x - 3 = 12$$

Check

$$A = \frac{1}{2}h(b_1 + b_2)$$

$A = 87$, $h = 6$, $b_1 = 17$, $b_2 = 12$:

$$87 = \frac{1}{2}(6)(17 + 12)$$
$$87 = 3(29)$$
$$87 = 87 \checkmark$$

Answer

$$x = 15$$

3a. The area of a trapezoid is 99. If the upper base is 13, the height is 6, and the lower base is represented by $4x$, find x.

3b. The shorter base of a trapezoid is represented by $3x + 2$ and the height is 7. If the area is 105 and the longer base is 19, find x.

Test

Area of a Rectangle = (Base)(Height)

Area of a Parallelogram = (Base)(Height)

Area of a Square = Side^2

Area of a Triangle = $\frac{1}{2}$(Base)(Height)

Area of a Circle = π(Radius)^2

Area of a Trapezoid = $\frac{1}{2}$(Height)(Upper base + Lower base)

1. Miriam is going to wallpaper her wall. Wallpaper costs $0.87 per square foot. If the wall measures 19 feet by 12 feet, how much will it cost to paper her wall?

2. The area around a pool is to be tiled. If the pool measures 45 feet by 32 feet and the tiled area around the pool measures 49 feet by 37 feet, how many square feet of tile are needed?

3. The length of a rectangular art gallery is 5 feet more than its width. The width is 35 feet. If the cost of sanding the floor is $1.26 per square foot, how much does it cost to sand the floor?

4. Wallpaper costs $2.16 per square *yard*. If a wall measures 12 *feet* by 24 *feet*, how much will it cost to wallpaper the entire wall?

5. The length of a rectangular living room is represented by $3x$ feet and the width is 16 feet. If its area is 384 square feet, find x.

6. The side of a square floor measures 26 feet. If we want to have the floor shellacked and the job costs $1.14 per square foot, what is the total cost?

7. Find a side of a square if its area is 9,409.

8. A side of a square is 15. If the area is represented by $9x - 36$, find x.

9. A piece of property is in the shape of a parallelogram. Its base is 84 feet, and its height is 67 feet. If the property sells for $5.22 a square foot, find the total cost of the property.

10. Parallelogram ABCD is 272 square feet in area. If its height is 17, find its base.

11. Ms. Robinson wants to sod a triangular area. The base is 64 feet and its height is 78 feet. Find the total cost of the sod if it sells for $0.56 a square foot.

12. If the area of a triangle is 810 and its height is 45, find its base.

13. The area of a triangle is 207. If the base is 18 and the height is represented by $3x + 2$, find x.

14. The base of a rectangle is 17, and its height is 11. The base of a triangle is 19, and its height is 24. Which figure is larger? By how much?

15. The radius of a circular tent is 7 feet. Find its area. Let $\pi = 3.14$.

16. The area of a circle is 200.96. Let $\pi = 3.14$. Find the radius of the circle.

17. A circle of plastic is to be cut from a large plastic sheet. If the radius of the circle is 5 feet and the plastic is a square 10 feet on a side, how much material is wasted (the cross-hatched area)? Let $\pi = 3.14$.

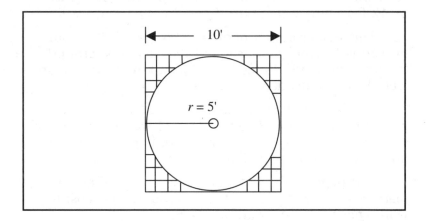

18. Find the area of a trapezoid with a lower base of 16, an upper base of 13, and a height of 9.

19. The area of a trapezoid is 497. Its height is 14, and its upper base is 32. Find its lower base.

20. The longer base of a trapezoid is represented by $2x + 4$, and the height is 16. If the area is 256 and the shorter base is 12, find x.

5
Volumes

A great deal of present-day two- and three-dimensional geometry still depends upon the propositions developed by the Greek mathematician Euclid (306–283 B.C.E.). His most significant contribution to geometry is contained in his 13 books of the *Elements*. Next to the Bible, the *Elements* is probably the most widely distributed and studied book in the world. The first four books cover plane geometry. The fifth and sixth volumes include the theory of proportions and similarity, while books seven through nine discuss number theory—prime numbers, divisibility of integers, etc. The last four books discuss solid geometry, and it is precisely in this area that we are going to discuss the formulas originally developed by Euclid.

5.1 Rectangular Solids

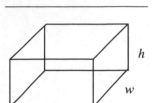

We can determine the volume of a rectangular solid by multiplying length, l, by width, w, by height, h.

$$V = lwh$$

where V = the volume of the rectangular solid, l = the length of the rectangular base, w = the width of the rectangular base, and h = the height of the rectangular solid. Volume is indicated in *cubic* dimensions because volume takes up the three dimensions of length, width, and height.

Example 1

Find the volume of a carton 3.5 meters long, 2 meters wide, and 4.7 meters high.

Analysis

Let V = the volume of the rectangular solid.

Let l = the length of the rectangular solid.

Let w = the width of the rectangular solid.

Let h = the height of the rectangular solid.

Use the volume formula $V = lwh$ and substitute.

Work

Volume formula: $\qquad\qquad V = lwh$

$l = 3.5$, $w = 2$, $h = 4.7$: $\qquad V = (3.5)(2)(4.7)$

$\qquad\qquad\qquad\qquad\qquad V = 32.9$

Answer

32.9 cubic meters

Exercises

$V_{\text{rectangular solid}} = lwh$

1a. Find the volume of a room 26 feet long, 15 feet wide, and 13 feet high.

1b. Melissa has a crate that is 5.3 meters wide, 3.2 meters high, and 6.7 meters long. Find its volume.

Example 2

The Convenience Corporation manufactures microwave ovens. The microwave oven is in the shape of a rectangular solid. It is 18.6 inches long and 9.4 inches high and has a volume of 1,398.72 cubic inches. Find its width.

Analysis

Let V = the volume of the microwave oven.

Let l = the length of the microwave oven.

Let w = the width of the microwave oven.

Let h = the height of the microwave oven.

Use the volume formula $V = lwh$ and substitute.

Work

$$V = lwh$$

$V = 1,398.72,$
$l = 18.6, h = 9.4:$

$$1,398.72 = (18.6)(w)(9.4)$$

$$1,398.72 = 174.84w$$

Divide by 174.84: $\qquad 8 = w$

Symmetric Property: $\qquad w = 8$

Answer

8 inches

2a. A fish tank is 18 inches long and 12 inches wide. If its volume is 2,505.6 cubic inches, find its height.

2b. The volume of a room is 4,118.4 cubic feet. If its length is 23.4 feet and its height is 11 feet, find its width.

Exercises

Example 3

Each student in a dormitory room should have 240 cubic feet of space. If the room measures 16 feet long by 20 feet wide by 12 feet high, what is the maximum number of students the room can accommodate?

Analysis

Let V = the volume of the room.

Let l = the length of the room.

Let w = the width of the room.

Let h = the height of the room.

Find the volume of the dormitory room. Then, since each pupil should have 240 cubic feet of space, divide the total volume by 240.

Work

$$V = lwh$$

$l = 16,\ w = 20,\ h = 12$: $V = 16 \times 20 \times 12$

$$V = 3{,}840$$

Divide the number of cubic feet of space in the room by 240, the requisite space for each student:

$$\frac{3{,}840}{240} = 16$$

Answer

16 students

Exercises

3a. Courie is pumping oxygen into an emergency room at the rate of 3 cubic meters per minute. If a room is 9 meters by 6 meters by 4 meters, how long will it take for the room to be filled with oxygen?

3b. A gasoline truck fills a storage tank at the rate of 72 cubic feet per minute. If the storage tank is in the shape of a rectangular solid and is 25 feet long by 16 feet wide by 9 feet high, how long will it take for an empty tank to be filled?

Example 4

A solid bar 4 feet long, 5 inches wide, and 8 inches high weighs 0.3 ounce per cubic inch. How much does the bar weigh?

Analysis

Let V = the volume of the rectanguar solid.

Let l = the length of the rectangular solid.

Let w = the width of the rectangular solid.

Let h = the height of the rectangular solid.

Change 4 feet to inches. Then find the volume in cubic inches and then multiply by 0.3, the number of ounces per cubic inch.

Work

$$V = lwh$$

$l = 4' = 48''$,
$w = 5''$, $h = 8''$: $\qquad V = 48 \times 5 \times 8 = 1{,}920 \, \text{cu in.}$

$0.3\,\text{oz/cu in.}$: $\qquad \text{weight} = 1{,}920 \times 0.3 = 576$

Answer

576 ounces or 36 pounds

Exercises

4a. A wooden plank weighs 0.02 ounce per cubic inch. If the plank is 6 feet long by 4 feet wide by 3 inches high, how much does it weigh?

4b. The metal door to Mikhail's apartment is 3 feet wide, 8 feet high, and 1.5 inches thick. If the metal weighs 0.3 ounce per cubic inch, how much does the door weigh?

Example 5

A plate of glass 4 feet 5 inches long and 3 feet 2 inches wide weighs 251.75 ounces. If the glass weighs 0.25 ounce per cubic inch, how thick is the glass (the height)?

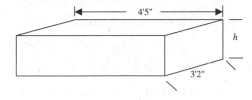

Analysis

Let V = volume, l = length, w = width, h = height. Since there is 0.25 ounce per cubic inch, we can just multiply the volume by 0.25 (oz/cu in.) to obtain the weight. Then, we'll substitute the given values to find the missing height or thickness. Change all dimensions to a common measure, inches.

Work

$$(0.25)V = \text{Weight}$$

$V = lwh$, Weight = 251.75: $\qquad (0.25)lwh = 251.75$

$l = 4'5'' = 53''$,
$w = 3'2'' = 38''$: $\qquad (0.25)(53)(38)h = 251.75$

$$503.5\,h = 251.75$$

Divide by 503.5: $\qquad\qquad\qquad h = 0.5$

Answer

0.5" thick

Exercises

5a. A sheet of plastic weighs 16.128 ounces. If it is 7 feet long by 4 feet wide and weighs 0.02 ounce per cubic inch, what is its thickness?

5b. The science lab has a metal plate that weighs 758.16 ounces. If it is 0.6 inch thick and 26 inches wide and weighs 0.9 ounce per cubic inch, what is its height?

Example 6

A rectangular solid is 17 inches long and 9 inches wide. If its volume is 1,989 cubic inches and its height is represented by $2x + 1$, find x.

Analysis

Let V = the rectangular solid.

Let l = the length of the rectangular solid.

Let w = the width of the rectangular solid.

Let h = the height of the rectangular solid.

Use the volume formula $V = lwh$ and substitute.

Work

$$V = lwh$$

$V = 1{,}989,\ l = 17,\ w = 9,$
$h = 2x + 1$:

$$1{,}989 = (17)(9)(2x + 1)$$

$$1{,}989 = 153(2x + 1)$$

$$1{,}989 = 306x + 153$$

Subtract 153: $\qquad 1{,}836 = 306x$

Divide by 306: $\qquad 6 = x$

Symmetric Property: $\qquad x = 6$

Answer

$$x = 6$$

Exercises

6a. A carton is 18 inches long and 16 inches wide. If its volume is 4,032 cubic inches and the height is represented by $3x - 1$, find x.

6b. Maria sleeps in a room that has a volume of 2,310 cubic feet. If it is 15 feet long by 14 feet wide and the height is represented by $3x + 2$, find x.

5.2 Cubes

A **cube** is simply a rectangular solid whose length, width, and height are all the same measure. Since all the dimensions are equal, let's call each dimension e, for edge.

$$\text{Volume} = e \cdot e \cdot e = e^3$$

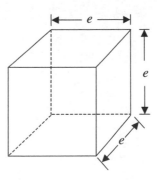

Example 1

To the nearest cubic inch, find the volume of cube container of gasoline if each edge is 13.6 inches. Round to the nearest cubic inch.

Analysis

Let $V =$ the volume of the cube.

Let $e =$ an edge of the cube.

Use the volume formula $V = e^3$ and substitute.

Work

$$V = e^3$$

$e = 13.6:$ $V = (13.6)^3$

$V = 2{,}515.456 \approx 2{,}515$

Answer

2,515 cubic inches

$V_{\text{cube}} = e^3$

1a. Find the volume of a cube whose edge is 34 centimeters.

1b. If one edge of a cube is 26 inches, find its volume.

Example 2

The weight of glycerin is 79.2 lb/ft³. Find the weight of a cube of glycerin whose edge is 6 inches. Round to the nearest tenth.

Analysis

Let $V =$ the volume of the cube.

Let $e =$ an edge of the cube.

Change 6 inches to feet, find the volume (in cubic feet), and then multiply by 79.2.

Work

$$V = e^3$$

$e = 6\,\text{in.} = 6/12\,\text{ft} = 0.5\,\text{ft}:$ $V = (0.5)^3$

$V = 0.125$ cubic foot

Multiply by 79.2:

$$(79.2)(0.125) = 9.9$$

Answer

9.9 pounds

2a. If concrete weighs 9.6 kilograms per cubic meter, find the weight of a cube of concrete whose edge is 2 meters.

2b. Metal screws weigh 3 kilograms per cubic foot. The screws were packaged in a cube whose edge was 1.5 feet. Find the weight of the cube. Round your answer to the nearest tenth of a kilogram.

Example 3

The volume of a cube is 512 cubic inches. Find one edge of the cube.

Analysis

Let V = the volume of the cube.

Let e = the edge of the cube.

Use the volume formula $V = e^3$ and substitute.

Work

$$V = e^3$$

$V = 512$: $\qquad\qquad\qquad 512 = e^3$

Find the cube root of 512: $\qquad 8 = e$

Symmetric Property: $\qquad\qquad e = 8$

Answer

8 inches

3a. If the volume of a cube is 0.512 cubic meter, find an edge.

3b. Find an edge of a cube if its volume is 343 cubic feet.

To obtain the volume of a triangular prism, we'll get the area of one triangle and multiply that by the height of the prism.

$$V = \frac{1}{2}(b)(h_t)(h_p)$$

where V = volume of prism, b = base of triangle, h_t = height of triangle, and h_p = height of prism. (Depending upon the way we view the prism, the height of the prism may be considered its length.) And since volume is in three dimensions, our measure is in cubic units.

Exercises

5.3 Prisms

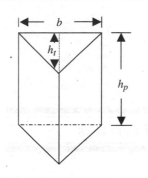

Example 1

One type of tent is in the shape of a prism. If the tent is 5.5 feet high, 17 feet long, and 8.4 feet wide, what is its volume?

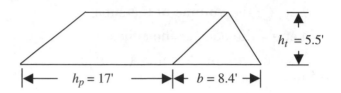

$h_t = 5.5'$

$h_p = 17'$ $b = 8.4'$

Analysis

Since this is a tent, we'll have to view this prism differently from our earlier prism. Now, the length of the tent is the same as h_p, the height of the other prism. Similarly, the width of the tent in our current example is the same as b, the base of the triangle in the prism just discussed.

Let V = the volume of the prism.

Let b = the base of the triangle.

Let h_p = the height or length of the prism.

Let h_t = the height of the triangle.

Work

Volume:

$$V = \frac{1}{2}(b)(h_t)(h_p)$$

$b = 8.4$, $h_t = 5.5$, $h_p = 17$: $V = \frac{1}{2}(8.4)(5.5)(17)$

$$V = 392.7$$

Answer

392.7 cu ft

Exercises

$V_{\text{prism}} = \frac{1}{2}(b)(h_t)(h_p)$

1a. The base of the triangular edge of a glass prism is 4.2 inches and its height is 2. If the length of the prism is 5 inches, find its volume.

1b. The base of the triangular edge of a prism is 3 inches. The height of the triangular edge of the prism is 7.8 inches, and the height of the prism is 13 inches. Find its volume.

Example 2

Bob and Harry's Ice Cream Company is going to market its ice cream in a new package—a triangular prism. The previous package was a rectangular solid. The triangular side of the new package has a base of 4.5 inches and a height of 5 inches; the entire prism is 6 inches high. The older rectangular solid was 5 inches long, 3.5 inches wide, and 4 inches high. Which solid contains less ice cream? How much less?

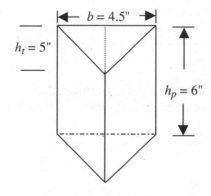

Analysis

Let V_s = the volume of the rectangular solid.

Let l = the length of the rectangular solid.

Let w = the width of the rectangular solid.

Let h = the height of the rectangular solid.

Let V_p = the volume of the prism.

Let b = the base of the triangle.

Let h_p = the height of the prism.

Let h_t = the height of the triangle.

Find the volume of the triangular prism and then find the volume of the rectangular solid. Compare the two volumes.

Work

Volume of prism:

$$V_p = \frac{1}{2}(b)(h_t)(h_p)$$

$b = 4.5$, $h_t = 5$, $h_p = 6$:

$$V_p = 0.5(4.5)(5)(6)$$

$$V_p = 67.5 \text{ cu in.}$$

Volume of

rectangular solid, V_s: $V_s = lwh$

$l = 5$, $w = 3.5$, $h = 4$: $V_s = (5)(3.5)(4)$

$V_s = 70\,\mathrm{cu\,in.}$

Difference $= 70 - 67.5 = 2.5$

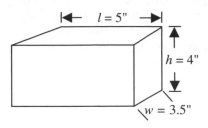

Answer

   The triangular prism contains 2.5 cubic inches less ice cream than the rectangular solid.

Exercises

$V_{\text{prism}} = \frac{1}{2}(b)(h_t)(h_p)$

2a. The ChocoLat Candy Company is going to begin a public relations campaign to promote their own candy bar against their competitor, the Pico Candy Company. ChocoLat markets their own candy bar in the shape of a prism while Pico markets a rectangular solid bar. As part of their public relations campaign, ChocoLat wants to advertise that "there is more chocolate" in their candy bar. The triangular edge of ChocoLat's bar has a height of 0.75 inch and a base of 1.2 inches. The entire prism is 5.4 inches high. Pico's bar in the shape of a rectangular solid is 1.7 inches wide, 4 inches long, and 0.4 inch high. Which bar has more chocolate? How much more?

2b. Nantucket Architecture, Inc., manufactures portable structures. One type of structure is in the shape of a rectangular solid and is 8 feet long, 8.3 feet high, and 9 feet wide. The second major type of structure is a prism. The base of its triangular edge is 9 feet and its height is 7.8 feet. The entire prism is 19 feet long. Which structure has a larger volume, and how much larger?

Example 3

An A-frame house is in the shape of a triangular prism. If the doorway is 5 feet wide, the house is 30 feet long, and the inside volume is 468.75 cubic feet, how high is the doorway?

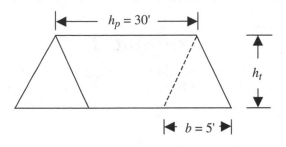

Analysis

Let V = the volume of the prism.

Let b = the base of the triangle.

Let h_p = the height of the prism.

Let h_t = the height of the triangle.

If we stand the prism on its side, we can see that its height of the prism is 30 feet, so $h_p = 30$, and the base of the triangle is 5 feet. We want to find the height of the triangle, h_t.

Work

$$V = \left(\frac{1}{2}\right)(bh_t)(h_p)$$

$V = 468.75$, $b = 5$,
$h_p = 30$:

$$468.75 = (0.5)(5)(h_t)(30)$$

$$468.75 = 75h_t$$

Divide by 75:

$$6.25 = h_t$$

Symmetric Property:

$$h_t = 6.25$$

Answer

6.25 ft high or $6\frac{1}{4}$ ft or 6 ft 3 in.

$V_{\text{prism}} = \frac{1}{2}(b)(h_t)(h_p)$

3a. The volume of a prism is 931 cubic meters. If the base of its triangular edge is 7 meters and its height is 14 meters, find the length of the prism.

3b. A shipping container is in the shape of a prism. If the height of its triangular edge is 6.4 feet, the length of the prism is 14 feet, and its volume is 492.8 cubic feet, find the base of its triangular edge.

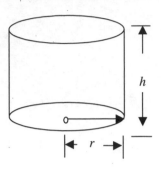

To find the volume of a cylinder, we'll just take the area of the base circle, πr^2, and multiply it by the height of the cylinder, h.

$$V = \pi r^2 h$$

where V = volume of cylinder, r = radius of circle, h = height of cylinder, and $\pi = 3.14$ or $\frac{22}{7}$.

Example 1

Chemicals are often shipped in cylinders. Find the volume of liquid contained in a cylinder 6 meters high with a radius of 3 meters. Let $\pi = 3.14$.

Analysis

Let V = the volume of the cylinder.

Let r = the radius of the circle.

Let h = the height of the cylinder.

Just use the formula for the volume of a cylinder. The answer is in cubic meters.

Work

$$V = \pi r^2 h$$

$\pi = 3.14$, $r = 3$, $h = 6$:
$$V = (3.14)(3)^2(6)$$
$$= (3.14)(9)(6) = 169.56$$

Answer

169.56 cubic meters

$V_{\text{cylinder}} = \pi r^2 h$

1–2. Let $\pi = 3.14$.

1a. A cylinder has a radius of 3 meters and a height of 6.4 meters. Find its volume.

1b. A cylindrical railroad car contains compressed hydrogen. Its radius is 5 feet and its length is 52 feet. Find its volume.

Example 2

A soda can is in the shape of a cylinder. If it contains 854.08 cubic centimeters of liquid and its height is 17 centimeters, find its radius. Let $\pi = 3.14$.

Analysis

Let V = the volume of the cylinder.

Let r = the radius of the circle.

Let h = the height of the cylinder.

Use the volume formula and substitute the given values.

Work

$$V = \pi r^2 h$$

$V = 854.08$, $\pi = 3.14$,
$h = 17$:

$$854.08 = (3.14)(r^2)(17)$$
$$854.08 = 53.38r^2$$

Divide by 53.38:

$$16 = r^2$$
$$4 = r$$

Symmetric Property:

$$r = 4$$

Answer

Radius = 4 centimeters

$V_{\text{cylinder}} = \pi r^2 h$

2a.   A cylindrical container holds 653.12 cubic feet of liquid. If its height is 13 feet, find its *radius*.

2b.   A cylinder is 9 feet long and has a volume of 706.5 cubic feet. Find its *diameter*.

Example 3

The Rockland Water Company is going to ship large cylindrical containers of water cross-country and is then going to repackage the water in smaller cube containers. A large cylindrical container has a radius of 3 feet and a height of 7 feet. The smaller cubes are 6 inches on each side. How many smaller containers can the large cylinder fill? Let $\pi = \frac{22}{7}$ and round the answer to the nearest whole number.

Analysis

Let $V_{cylinder}$ = the volume of the cylinder.

Let $\pi = \frac{22}{7}$.

Let r = the radius of the circle.

Let h = the height of the cylinder.

Let V_{cube} = the volume of the cube.

Let e = the edge of the cube.

First, we'll find the volume of water in the large cylinder. Then, we'll find the volume in the smaller cube container. Finally, we'll divide in order to find how many smaller containers of water the large cylinder can fill.

Work

Volume of cylinder, $V_{cylinder}$:

$$V_{cylinder} = \pi r^2 h$$

$\pi = \frac{22}{7}, r = 3, h = 7$:

$$= \frac{22}{7}(3)^2(7)$$

$$= \frac{22}{7}(9)(7)$$

$$= 198 \text{ cubic feet}$$

There are 198 cubic *feet* in the large cylinder, but the cube is measured in cubic *inches*. To be consistent, let's change the measurement unit of the cube from inches to feet.

Volume of cube, V_{cube}:

$$V_{cube} = e^3$$

$e = 6 \text{ in.} = \frac{6}{12} \text{ft} = \frac{1}{2} \text{ft} = 0.5 \text{ ft}$:

$$V_{cube} = (0.5)(0.5)(0.5)$$

$$V_{cube} = 0.125 \text{ cu ft}$$

Now, to find out how many cubes can be filled by the large cylinder, we'll divide the volume of the large cylinder, $V_{cylinder}$, by the volume of one small cube, V_{cube}:

$$\frac{V_{cylinder}}{V_{cube}} = \frac{198}{0.125} = 1,584$$

Answer

1,584 containers

$$V_{cylinder} = \pi r^2 h$$

Let $\pi = \frac{22}{7}$.

3a.   A cylinder has a 3-foot radius and is 7 feet high. A rectangular solid is 5 feet wide, 6 feet long, and 7 feet high. Which is larger? By how much? Round to the nearest cubic foot.

3b. The Krown Soft Drink Company wants to ship its soft drinks in a large cylinder and then repackage the drinks in individual prisms. The cylinders have a radius of 3 feet and are 6 feet tall. The prisms each have a triangular base of 15 inches and a height of 18 inches, while the prism itself is 12 inches tall. How many prisms can be filled with drinks from one large cylinder? Round your answer to the *lowest* whole number.

The volume of a cone is one third the volume of a cylinder with the same base and height.

$$V = \frac{1}{3}\pi r^2 h$$

where V = volume of cone, r = radius of base circle, and h = height of cone.

5.5 Cones

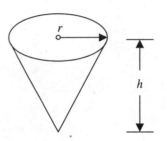

Example 1

The diameter of a large cone is 3 feet and its height is 7.5 feet. Find its volume. Round to the nearest hundredth of a cubic foot and let $\pi = \frac{22}{7}$.

Analysis

Let V = the volume of the cone.

Let r = the radius of the circle.

Let h = the height of the cone.

First, find the radius of the cone. Then, use the formula for the volume and just substitute.

Work

Radius = $\frac{1}{2}$ × Diameter: $r = (0.5)(3) = 1.5$

Volume: $V = \frac{1}{3}\pi r^2 h$

$\pi = \frac{22}{7}$, $r = 1.5$, $h = 7.5$: $V = \frac{1}{3} \times \frac{22}{7}(1.5)^2(7.5)$

$V = 17.67857 \approx 17.68$

Answer

17.68 cu ft

$V_{cone} = \frac{1}{3}\pi r^2 h$

Let $\pi = \frac{22}{7}$.

1a. A cone has a radius of 3 feet and is 14 feet tall. Find its volume.

1b. The diameter of a cone is 14 inches and its height is 6 inches. Find its volume.

Example 2

A cone-shaped storage tank can hold 462 cubic feet of liquid. If the storage tank is 9 feet high, find the *diameter* of the base. Let $\pi = \frac{22}{7}$.

Analysis

Let V = the volume of the cone.

Let $\pi = \frac{22}{7}$.

Let r = the radius of the circle.

Let h = the height of the cone.

Use the volume formula to find the radius. Then double it to find the diameter.

Work

$$V = \frac{1}{3}\pi r^2 h$$

$V = 462, \pi = \frac{22}{7}, h = 9{:}$
$$462 = \frac{1}{3}\left(\frac{22}{7}\right)r^2(9)$$

$$462 = \frac{198}{21}r^2$$

Multiply by $\frac{21}{198}$:
$$\frac{21}{198}\left(462 = \frac{198}{21}r^2\right)$$

$$\frac{9{,}702}{198} = r^2$$

$$49 = r^2$$

Take the square root:
$$7 = r$$

Symmetric Property:
$$r = 7$$

Diameter = 2 × Radius:
$$D = (2)(7) = 14$$

Answer

Diameter = 14 feet

$V_{cone} = \frac{1}{3}\pi r^2 h$

2a.   A cone-shaped container can hold 1,188 cubic feet of liquid. If its height is 14 feet, find its radius.

2b.   The diameter of a cone is 12 feet. If the cone has a volume of 792 cubic feet, find its height.

Example 3

A silo is shaped like a cone and contains wheat. The radius is 10 feet and the height is 15 feet. If the silo can release wheat from its bottom at the rate of 25 cubic feet per minute, how long would it take for the silo to empty fully? Round the answer to the nearest minute and let $\pi = 3.14$.

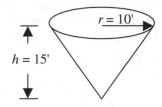

Analysis

Let V = the volume of the cone.

Let r = the radius of the circle.

Let h = the height of the cone.

First, find the volume of the silo. The wheat can be released at the rate of 25 cubic feet per minute, so after we obtain the volume, we'll divide by 25 to find the time it takes to entirely empty the silo.

Work

$$V = \frac{1}{3}\pi r^2 h$$

$\pi = 3.14, r = 10, h = 15$: $V = \frac{1}{3}(3.14)(10)^2(15)$

$$V = \frac{1}{3}(3.14)(100)(15)$$

$$V = 1,570 \text{ cubic feet}$$

Now we'll divide the volume, 1,570 cubic feet, by the amount of wheat that can be released per minute, 25 cubic feet. The answer represents the amount of time it takes for the silo to empty.

$$\frac{1{,}570 \text{ cu ft}}{25 \text{ cu ft/min}} = 62.8 \approx 63$$

Answer

   63 minutes

Exercises

$V_{\text{cone}} = \frac{1}{3}\pi r^2 h$

Let $\pi = 3.14$.

3a.   An ice-cream cone is 6 inches tall and has a radius of 2 inches. If the cone is just filled up to the brim and not higher and the ice-cream drips out of a hole in the bottom at the rate of 0.2 cubic inch per second, how long will it take for the cone to empty out? Round to the nearest second.

3b.   Oil flows out of an industrial cone at the rate of 0.5 cubic feet per second. If the cone has a 5-foot radius and is 9 feet tall, how many seconds will it take for the cone to empty? Round the answer to the nearest second.

5.6 Spheres

The volume, V, of a sphere equals $\frac{4}{3}\pi$ times the cube of the radius, r:

$$V = \frac{4}{3}\pi r^3$$

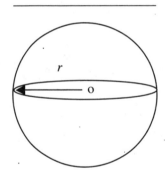

Example 1

A sphere has a diameter of 9 centimeters. Find its volume. Round to the nearest tenth of a cubic centimeter. Let $\pi = 3.14$.

Analysis

   Let V = the volume of the sphere.

   Let r = the radius of the sphere.

   First, find the radius and then substitute in the formula for volume.

Work

Radius $= \frac{1}{2} \times$ Diameter: Radius $= \frac{1}{2}(9) = 4.5$ cm

Volume: $V = \left(\frac{4}{3}\right)\pi r^3$

$\pi = 3.14$, $r = 4.5$: $V = (1.333)(3.14)(4.5)^3$

$V = 381.41462$

Answer

   381.4 cu cm

$V_{\text{sphere}} = \frac{4}{3}\pi r^3$

Let $\pi = 3.14.$

1a. A balloon has a radius of 3 meters. Find its volume. Round to the nearest tenth of a cubic meter.

1b. The radius of a basketball is 4 inches. Find its volume. Round to the nearest cubic inch.

Example 2

A solid metal sphere has a volume of 2,376/21 cubic inches. Find its radius. Let $\pi = \frac{22}{7}$.

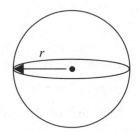

Analysis

Let V = the volume of the sphere.

Let r = the radius of the sphere.

Just use the formula and substitute 2,376/21 for the volume.

Work

Volume formula:
$$V = \frac{4}{3}\pi r^3$$

$V = \frac{2,376}{21}$, $\pi = \frac{22}{7}$:
$$\frac{2,376}{21} = \frac{4}{3} \cdot \frac{22}{7} r^3$$

$$\frac{2,376}{21} = \frac{88}{21} r^3$$

Multiply by $\frac{21}{88}$:
$$\frac{21}{88}\left(\frac{2,376}{21} = \frac{88}{21} r^3\right)$$

$$27 = r^3$$

Take the cube root of 27: $3 = r$

Symmetric Property: $r = 3$

Answer

3 inches

$V_{\text{sphere}} = \frac{4}{3}\pi r^3$

Let $\pi = \frac{22}{7}$.

2a. The volume of a sphere is 704/21. Find its radius.

2b. A spherical balloon has a volume of 11,000/21 cubic inches. Find its radius.

Example 3

A hot air spherical balloon has a diameter of 6 feet. If the mechanical air pump can fill the balloon at the rate of 30 cubic feet per minute, how long will it take to fill the balloon? Round to the nearest minute and let $\pi = \frac{22}{7}$.

Analysis

Let V = volume of the balloon.

Let r = the radius of the balloon.

Let d = the diameter of the balloon.

First, find the radius. Then determine the volume of the balloon. Finally, divide by 30 cu ft/min to determine the time it takes to fill the balloon.

Work

$$r = \frac{1}{2}d$$

$d = 6$: $\qquad r = \frac{1}{2}6 = 3$

$$V = \frac{4}{3}\pi r^3$$

$r = 3, \pi = \frac{22}{7}$: $\qquad V = \frac{4}{3} \cdot \frac{22}{7} \cdot 3^3$

$$V = 113.14285$$

The balloon's volume is 113.14285 cu ft. The air pump fills it at the rate of 30 cubic feet per minute, so we'll divide:

$$\frac{113.14285}{30} = 3.7714283 \approx 4$$

Answer

4 minutes

3a. A metal sphere contains liquid nitrogen. If the sphere has a radius of 3 feet and the nitrogen is discharged at the rate of 2 cubic feet per second, how many seconds will it take to empty the sphere? Round to the nearest second.

3b. A spherical balloon with a diameter of 4 feet leaks gas at the rate of 0.4 cubic feet per second. In how many seconds will the balloon be completely deflated? Round the answer to the nearest second.

Test

Volume of a rectangular solid:	$V = lwh$, where V = volume, l = length, w = width, and h = height.
Volume of a cube:	$V = e^3$, where V = volume and e = edge.
Volume of a prism:	$V = \frac{1}{2}(b)(h_t)(h_p)$, where V = volume, b = base of triangle, h_t = height of the triangle, and h_p = height of the prism.
Volume of a cylinder:	$V = \pi r^2 h$, where V = volume, π = 3.14 or $\frac{22}{7}$, r = radius, and h = height.
Volume of a cone:	$V = \frac{1}{3}\pi r^2 h$, where V = volume, π = 3.14 or $\frac{22}{7}$, r = radius, and h = height.
Volume of a sphere:	$V = \frac{4}{3}\pi r^2 h$, where V = volume, π = 3.14 or $\frac{22}{7}$, r = radius.

1. A fish tank is 18 inches long, 12.6 inches wide, and 17 inches high. What is its volume?

2. A packing crate is 22 inches long and 14 inches wide. If its volume is 2,556.4 cubic inches, how high is the crate?

3. Liquid soap is being poured into a rectangular-shaped container at the rate of 400 cubic inches per minute. If the container measures 25 inches long by 14 inches wide by 36 inches high, how long will it take to fill the container?

4. Each cubic centimeter of a glass plate weighs 0.3 ounce. If the glass plate measures 40 centimeters long by 35 centimeters wide by 6 centimeters thick, how much does the plate weigh?

5. A wooden board weighs 2,304 ounces. If it is 5 feet long by 4 feet wide and weighs 0.4 ounce per cubic inch, what is its thickness?

6. Find the volume of a cube whose edge is 16 centimeters.

7. If alcohol weighs 2.4 kilograms per cubic meter, find the weight of a cube container of alcohol whose edge is 3 meters.

8. If the volume of a cube is 0.729 cubic meter, find an edge.

9. A tent in the shape of a prism is 13 feet long. Its triangular base is 6 feet wide and 5.9 feet high. Find its volume.

10. Felix candy bars are manufactured in the shape of a triangular prism. The height of its triangular edge is 0.7 inch and the base is 0.6 inch. The prism is 5 inches long. Krispy candy bars are marketed in the shape of a rectangular solid and measure 4.6 inches long, 0.5 inch thick, and 0.4 inch wide. Which candy bar has a greater volume? How much greater?

11. The volume of a triangular prism is 2,380 cubic meters. If its length is 14 meters and the base of its triangular edge is 17 meters, find the height of the triangular edge.

12. A cylinder is 17 meters long and has a volume of 854.08 cubic meters. Find its radius. Let $\pi = 3.14$.

13. If the diameter of a cone is 8 inches and its height is 21 inches, find its volume. Let $\pi = \frac{22}{7}$.

14. A cone-shaped container can hold 132 cubic feet of liquid. If its height is 14 feet, find its radius. Let $\pi = \frac{22}{7}$.

15. A loading container is in the shape of a cone. It has a diameter of 4 feet and is 10 feet tall. If it can unload grain at the rate of 3 cubic feet per second, how long will it take to unload the entire cone? Let $\pi = 3.14$ and round your answer to the nearest second.

16. The volume of a cylinder is 1,808.64 cubic meters. If its radius is 8 meters, find its height. Let $\pi = 3.14$.

17. A cylinder contains ice cream. Its radius is 5 inches and it stands 12 inches tall. The Cool Man Ice Cream Shop sells ice-cream cones. If a cone is 5 inches tall and has a diameter of 3 inches, to the nearest whole number, how many ice-cream cones can be scooped out of the cylinder? Let $\pi = 3.14$.

18. A sphere has a radius of 4 meters. Find its volume. Let $\pi = 3.14$ and round to the nearest cubic meter.

19. The volume of a sphere is 5,632/21. Find its radius. Let $\pi = \frac{22}{7}$.

20. A spherical balloon with a diameter of 4 feet leaks gas at the rate of 0.6 cubic feet per second. In how many seconds will the balloon be completely deflated? Round off your answer to the nearest second and let $\pi = 3.14$.

6
Integers and Numbers

Cavepeople were the first mathematicians. They started counting their individual possessions such as apples, animals, etc.—what we call the set of "natural" or "counting numbers."

Eventually 0 was added to make the set of "whole numbers": {0, 1, 2, 3, . . .}.

The ancient civilizations of Mesopotamia, India, China, Egypt, and Babylon developed the idea of whole numbers and were able to perform what we consider standard operations—addition, subtraction, multiplication, and division. Whereas these peoples applied mathematics to practical situations, the Greeks were probably the first to develop the logical structure of mathematics. Thales of Miletus (*c.*636–*c.*546 B.C.E.)

and Pythagoras of Samos (*c*.560–*c*.480 B.C.E.) were *the* primary Greek philosopher/mathematicians and they introduced the idea of abstract mathematics. They thought of numbers as theoretical concepts and geometrical shapes as imaginative constructs rather than actual distances or real rectangles and circles.

Greek mathematicians in general and Pythagoras in particular believed in a rational universe wherein order took the place of random chaos. The School of Pythagoras investigated the relationship between whole numbers and their rational offspring and believed that this logical relationship underlay the structure of the universe.

6.1 Whole Numbers

Let's try to solve some algebraic whole number problems—the same type of problems which Pythagoras and Euclid solved.

Example 1

Six less than twice a whole number is 48. Find the number.

Analysis

Let x = the whole number.

Let $2x - 6$ = six less than twice a whole number.

Work

Six less than twice a number is 48: $2x - 6 = 48$

Add six: $2x = 54$

Divide by two: $x = 27$

Check

$$2x - 6 = 48$$

Let $x = 27$: $2(27) - 6 = 48$

$$54 - 6 = 48$$

$$48 = 48 \checkmark$$

Answer

27

Exercises

1a.   Five more than three times a whole number is 71. Find the number.

1b.   If 13 less than seven times a whole number is 204, find the number.

Example 2

One whole number is three times another. Together, the numbers add up to 84. Find both numbers.

Analysis

Let x = the first whole number.

Let $3x$ = the second whole number.

Work

The two numbers add up to 84:

$$x + 3x = 84$$
$$4x = 84$$

Divide by 4:

$$x = 21$$
$$3x = 63$$

Check

$$x + 3x = 84$$
$$21 + 63 = 84$$
$$84 = 84 \checkmark$$

Answer

21, 63

Exercises

2a.   One whole number is 4 times another. Together, the two numbers add up to 100. Find both numbers.

2b.   One whole number is 12 more than another. If the two numbers add up to 70, find both numbers.

Example 3

The larger of two whole numbers is five times the smaller. If the larger number is increased by 3, it equals 15 more than the smaller. Find both numbers.

Analysis

Let x = the smaller number.

Let $5x$ = the larger number.

Work

If the larger number is increased by 3, it equals 15 more than the smaller.

$$5x + 3 = x + 15$$

Subtract 3: $\qquad 5x = x + 12$

Subtract x: $\qquad 4x = 12$

Divide by 4: $\qquad x = 3$

$$5x = 15$$

Check

$$5x + 3 = x + 15$$

$$15 + 3 = 3 + 15$$

$$18 = 18 \checkmark$$

Answer

3, 15

Exercises

3a. One whole number is three times a second. If 20 is added to the smaller number, the result equals 6 more than the larger number. Find both numbers.

3b. The larger of two whole numbers is 7 more than the smaller. Double the smaller number equals 8 more than the larger number. Find both numbers.

When we say "separate a number into two parts," we expect you to find two numbers whose sum equals the original number. For example, when we say "separate 7 into two parts," there are a number of ways we can do just that:

One Part	Second Part
6	1
5	2
4	3
x	$7 - x$

Example 4

Separate 15 into two parts such that twice the smaller is 3 more than the larger.

Analysis

Let x = the smaller part.

Let $15 - x$ = the larger part.

Work

        Twice the smaller is three more than the larger:

$$2(x) = (15 - x) + 3$$
$$2x = 15 - x + 3$$
$$2x = 18 - x$$

Add x: $3x = 18$

Divide by 3: $x = 6$

$$15 - x = 9$$

Check

$$2x = (15 - x) + 3$$
$$2(6) = 9 + 3$$
$$12 = 12 \checkmark$$

Answer

    6, 9

4a.   Separate 29 into two parts such that three times the smaller is 19 more than the larger.

4b.   Separate 45 into two parts such that twice the smaller is 9 less than the larger.

    If we eliminate fractions and months, age problems are just a subset of integer problems.

Exercises

6.2 Age Problems

Example 1

    If Latisha is 5 years older than Wanda and Wanda is represented by $8a$ years, how would we represent Latisha's age?

Analysis

    Let $8a$ = Wanda's age.

    Just add 5 to that to designate Latisha's age.

Work

        Latisha's age: $8a + 5$

Answer

    $8a + 5$

1a. Vincent is twice as old as Frank. If Frank's age is represented by $3x$, how would you represent Vincent's age?

1b. Julia is 5 years younger than Margot. If Margot's age is represented by $6t + 7$, how would Julia's age be represented?

Example 2

Mickey is 7 years younger than Juan. Together, their ages add up to 19. Find both their ages.

Analysis

Let x = Juan's age.

Let $x - 7$ = Mickey's age.

Work

Their ages add up to 19: $x + (x - 7) = 19$

$$2x - 7 = 19$$

Add 7: $2x = 26$

Divide by 2: $x = 13$

$$x - 7 = 6$$

Check

$$x + (x - 7) = 19$$
$$13 + 6 = 19$$
$$19 = 19 ✓$$

Answer

Juan is 13; Mickey is 6.

2a. Heinrich is 12 years older than Malka. Together, their ages add up to 48. How old is each?

2b. Guillermo is twice as old as Jesus. Together, their ages add up to 42. How old is Guillermo?

Example 3

Darnell is 40 years old. If he is 4 years older than three times Maria's age, how old is Maria?

Analysis

Let x = Maria's age.

Let $3x + 4$ = Darnell's age.

Work

Darnell is 40:	$3x + 4 = 40$
Subtract 4:	$3x = 36$
Divide by 3:	$x = 12$

Check

$$3x + 4 = 40$$

$x = 12$:

$$3(12) + 4 = 40$$

$$36 + 4 = 40$$

$$40 = 40 \checkmark$$

Answer

12

Exercises

3a.   Pedro is 39 years old. He is 1 year less than twice Mike's age. How old is Mike?

3b.   Santos is 4 years older than three times David's age. If Santos is 28 years old, how old is David?

Example 4

Billy is twice Gustave's age. Li Mai is 5 years older than Gustave. If their combined ages equal 61, how old is each?

Analysis

Let x = Gustave's age.

Let $2x$ = Billy's age.

Let $x + 5$ = Li Mai's age.

Work

Their combined ages equal 61:

$$(x)+(2x)+(x+5) = 61$$

$$4x + 5 = 61$$

Subtract 5: $\qquad\qquad 4x = 56$

Divide by 4: $\qquad\qquad x = 14$ (Gustave's age)

$$2x = 28 \text{ (Billy's age)}$$

$$x+5 = 19 \text{ (Li Mai's age)}$$

Check

$$(x)+(2x)+(x+5) = 61$$

$x = 14, 2x = 28,$
$x + 5 = 19:$ $\qquad\qquad 14 + 28 + 19 = 61$

$$61 = 61 \checkmark$$

Answer

Gustave is 14, Billy is 28, and Li Mai is 19.

Exercises

4a.   Philomena is 5 years older than Judy. Mallory is 11 years younger than twice Judy's age. If their combined ages add up to 62, how old is Mallory?

4b.   Charles is 9 years older than Nestor. Marcus is 1 year younger than twice Nestor's age. Together, their ages add up to 80. How old is Marcus?

6.3 Consecutive Integers

The set of **integers** includes the positive and negative whole numbers as well as zero:

$$\{\ldots, -4, -3, -2, -1, 0, 1, 2, 3, 4, \ldots\}$$

Consecutive integers are integers that follow one another:

$$1, 2, 3, \ldots$$
$$56, 57, 58, \ldots$$
$$-7, -6, -5, \ldots$$

If we let x represent an integer, $x + 1$ represents the next consecutive integer, $x + 2$ the integer after that, and so forth:

$$x, x + 1, x + 2, \ldots$$

Let $x^2 + y$ represent an integer. Then, $(x^2 + y) + 1$ represents the next consecutive integer and $\{(x^2 + y) + 1\} + 1$ represents the integer after that and so forth:

$$x^2 + y, x^2 + y + 1, x^2 + y + 2, \ldots$$

Example 1

Find two consecutive integers whose sum is 197.

Analysis

Let x = the first integer.

Let $x + 1$ = the second integer.

Work

The sum of the two integers is 197.

$$x + (x + 1) = 197$$
$$x + x + 1 = 197$$
$$2x + 1 = 197$$

Subtract 1: $\qquad 2x = 196$

Divide by 2: $\qquad x = 98$

$$x + 1 = 99$$

Check

$$x + (x + 1) = 197$$
$$98 + 99 = 197$$
$$197 = 197 \checkmark$$

Answer

98, 99

Exercises

1a. Find two consecutive integers that add up to 109.

1b. Determine two consecutive integers whose sum is 223.

Example 2

Find three consecutive integers whose sum is 54.

Analysis

Let x = the first integer.

Let $x + 1$ = the second integer.

Let $x + 2$ = the third integer.

Work

The sum of the three
integers is 54:

$$x + (x + 1) + (x + 2) = 54$$
$$x + x + 1 + x + 2 = 54$$
$$3x + 3 = 54$$

Subtract 3:

$$3x = 51$$

Divide by 3:

$$x = 17$$
$$x + 1 = 18$$
$$x + 2 = 19$$

Check

$$x + (x + 1) + (x + 2) = 54$$
$$17 + 18 + 19 = 54$$
$$54 = 54 \checkmark$$

Answer

17, 18, 19

Exercises

2a.   If three consecutive integers add up to 225, what are they?

2b.   Four consecutive integers add up to 98. Find them.

Example 3

Determine three consecutive integers such that twice the sum of the first and second is 27 more than three times the third.

Analysis

Let x = the first integer.

Let $x + 1$ = the second integer.

Let $x + 2$ = the third integer.

Work

Twice the sum of the first and second is 27 more than three times the third:

$$2[x+(x+1)] = 3(x+2)+27$$
$$4x + 2 = 3x + 6 + 27$$
$$4x + 2 = 3x + 33$$

Subtract 2: $$4x = 3x + 31$$
Subtract 3x: $$x = 31$$
$$x + 1 = 32$$
$$x + 2 = 33$$

Check

$$2[x+(x+1)] = 3(x+2)+27$$
$$2(31+32) = 3(33)+27$$
$$2(63) = 99+27$$
$$126 = 126 \checkmark$$

Answer

31, 32, 33

3a.   Find three consecutive integers such that three times the largest is 9 less than four times the smallest.

3b.   What are three consecutive integers such that three times the smallest added to 7 is 13 less than four times the largest?

Exercises

Consecutive even integers are even integers that follow one another. We have to add 2 to the preceding even integer to get to the next consecutive even integer:

6.4 Consecutive Even Integers

$$4, 6, 8, \ldots$$
$$108, 110, 112, \ldots$$
$$-24, -22, -20, \ldots$$

If n represents the first even integer, $n + 2$ represents the next consecutive even integer, $n + 4$ the even integer after that, and so forth:

$$n, n + 2, n + 4, \ldots$$

If $n + 5$ represents the first even integer, $(n + 5) + 2$ and $(n + 5) + 4$ represent the next two consecutive even integers:

$$n+5, (n+5)+2, (n+5)+4 \qquad \text{or}$$
$$n+5, n+7, n+9$$

Example 1

If z represents the first even integer, find the next two consecutive even integers.

Analysis

Let z = the first even integer.

Add 2 to obtain the second even integer.

Add 4 to obtain the third even integer.

Work

$$z$$
$$z + 2$$
$$z + 4$$

Answer

$z + 2$, $z + 4$

1a. P represents the first even integer. Find the next three consecutive even integers.

1b. If $7t$ represents the first of three consecutive even integers, find the next two consecutive even integers.

Example 2

If $3m + 1$ represents the first even integer, what are the next two consecutive even integers?

Analysis

Let $3m + 1$ = the first even integer.

Keep adding 2's to obtain the next even integers.

Work

$$(3m + 1)$$
$$(3m + 1) + 2 = 3m + 3$$
$$(3m + 1) + 4 = 3m + 5$$

Answer

$3m + 3$, $3m + 5$

2a. $7x + 5$ represents the first of three consecutive even integers. Represent the next two consecutive even integers. Simplify your answers.

2b. If $3x^2 - 2$ represents the first even integer, find the next three consecutive even integers.

Exercises

Example 3

Find two consecutive positive even integers whose sum is 94.

Analysis

Let $x = $ the first even integer.

Let $x + 2 = $ second even integer.

Work

The sum of the integers is 94:

$$x + (x + 2) = 94$$
$$x + x + 2 = 94$$
$$2x + 2 = 94$$

Subtract 2: $\qquad 2x = 92$

Divide by 2: $\qquad x = 46$

$$x + 2 = 48$$

Check

$$x + (x + 2) = 94$$
$$46 + 48 = 94$$
$$94 = 94 \checkmark$$

Answer

46, 48

3a. Find three consecutive even integers whose sum is 132.

3b. Four consecutive even integers add up to 324. Find them.

Exercises

Example 4

Find three consecutive positive even integers such that three times the smallest is 16 more than the sum of the second and third consecutive even integers.

Analysis

Let x = the first even integer.

Let $x + 2$ = the second even integer.

Let $x + 4$ = the third even integer.

Work

Three times the smallest is 16 more than the sum of the second and third.

$$3(x) = (x+2)+(x+4)+16$$
$$3x = 2x + 22$$

Subtract 2x: $\quad\quad x = 22$

$$x + 2 = 24$$
$$x + 4 = 26$$

Check

$$3(x) = (x+2)+(x+4)+16$$
$$3(22) = 24+26+16$$
$$66 = 66 \checkmark$$

Answer

22, 24, 26

Exercises

4a. Find three consecutive even integers such that four times the smallest is 108 less than five times the largest integer.

4b. Find three consecutive even integers such that twice the second is 34 more than the largest integer.

Example 5

Janine, Melissa, and Jack live in consecutively even-numbered houses. If Jack's house number is 48 less than twice Janine's house number, find Melissa's house number.

Analysis

Let Janine's first even-numbered house = x.

Let Melissa's second even-numbered house = $x + 2$.

Let Jack's third even-numbered house = $x + 4$.

Work

Jack's house number is 48 less than twice Janine's house number.

$$x + 4 = 2(x) - 48$$
$$x + 4 = 2x - 48$$

Subtract x: $\quad 4 = x - 48$

Add 48: $\quad 52 = x$

Symmetric Property: $\quad x = 52\,(\text{Janine})$

$$x + 2 = 54\,(\text{Melissa})$$
$$x + 4 = 56\,(\text{Jack})$$

Check

$$x + 4 = 2(x) - 48$$
$$52 + 4 = 2(52) - 48$$
$$56 = 104 - 48$$
$$56 = 56 \checkmark$$

Answer

54

5a.   Julissa has three jars of coins, A, B, and C. The jars contain coins that are in consecutive even-number order. Three times the number of coins in jar B is 42 more than twice the number of coins in jar C. Find the number of coins in jar A.

5b.   The ages of Lee, Miriam, and Kendra are in consecutive even-number order. Miriam is 10 years younger than twice Lee's age. Find Kendra's age.

Exercises

6.5 Consecutive Odd Integers

Consecutive odd integers are odd integers that follow one another. Just as with even integers, we have to add 2 to the preceding odd integer in order to get to the next consecutive odd integer:

$$9, 11, 13, \ldots$$
$$97, 99, 101, \ldots$$
$$-15, -13, -11, \ldots$$

If $\frac{1}{2}x^2 + 5$ represents the first odd integer, $(\frac{1}{2}x^2 + 5) + 2$ represents the next consecutive odd integer and $(\frac{1}{2}x^2 + 5) + 4$ represents the odd integer after that:

$$\frac{1}{2}x^2 + 5, \left(\frac{1}{2}x^2 + 5\right) + 2, \left(\frac{1}{2}x^2 + 5\right) + 4 \qquad \text{or}$$

$$\frac{1}{2}x^2 + 5, \frac{1}{2}x^2 + 7, \frac{1}{2}x^2 + 9$$

Example 1

If p represents the first odd integer, find the next two consecutive odd integers.

Analysis

Let p = the first odd integer.

Keep adding 2's to obtain consecutive odd integers.

Work

$$p$$
$$p + 2$$
$$p + 4$$

Answer

$$p + 2, p + 4$$

1a. $6R + 1$ represents the first odd integer. Find the next odd integer.

1b.   If $4t^3 + 3$ represents the first odd integer, find the next odd integer.

Exercises

Example 2

If $j - 1$ represents the first odd integer, find the next three consecutive odd integers.

Analysis

Let $(j - 1)$ = the first odd integer.

Keep adding 2's to obtain consecutive odd integers.

Work

$$(j - 1)$$
$$(j - 1) + 2 = j + 1$$
$$(j - 1) + 4 = j + 3$$
$$(j - 1) + 6 = j + 5$$

Answer

$$j + 1, j + 3, j + 5$$

2a. $2x + 5$ represents the first odd integer. Find the next odd integer.

2b.   If $7t^2 + 4$ represents the first odd integer, find the next odd integer.

Exercises

Example 3

Find four consecutive odd integers whose sum is 328.

Analysis

Let x = the first odd integer.

Let $x + 2$ = the second odd integer.

Let $x + 4$ = the third odd integer.

Let $x + 6$ = the fourth odd integer.

Work

The sum of the four consecutive odd integers is 328.

$$x + (x + 2) + (x + 4) + (x + 6) = 328$$

$$4x + 12 = 328$$

Subtract 12: $\qquad\qquad\qquad\qquad 4x = 316$

Divide by four: $\qquad\qquad\qquad\qquad x = 79$

$$x + 2 = 81$$

$$x + 4 = 83$$

$$x + 6 = 85$$

Check

$$x + (x + 2) + (x + 4) + (x + 6) = 328$$

$$79 + 81 + 83 + 85 = 328$$

$$328 = 328 \checkmark$$

Answer

79, 81, 83, 85

Exercises

3a. Find three consecutive odd integers whose sum is 117.

3b. The sum of four consecutive odd integers is 288. Find them.

Example 4

Is it possible to find three consecutive odd integers such that twice the third is 57 less than three times the sum of the first and second? If so, what are the integers?

Analysis

Let x = the first odd integer.

Let $x + 2$ = the second odd integer.

Let $x + 4$ = the third odd integer.

Work

Twice the third integer is 57 less than three times the sum of the first and second.

$$2(x + 4) = 3(x + x + 2) - 57$$
$$2x + 8 = 3(2x + 2) - 57$$
$$2x + 8 = 6x + 6 - 57$$
$$2x + 8 = 6x - 51$$

Subtract 8: $\qquad 2x = 6x - 59$

Subtract 6x: $\qquad -4x = -59$

Divide by –4: $\qquad x = \dfrac{-59}{-4} = 14\dfrac{3}{4}$

Answer

$14\frac{3}{4}$ is not an *integer*, so it is not possible to find three consecutive odd integers to fulfill the conditions.

Exercises

4a.   Is it possible to find three consecutive odd integers such that three times the second is 77 less than four times the first? If so, what are they?

4b.   The sum of the first and fourth of four consecutive odd integers is 91 less than three times the second. Find all four integers.

Test

1–15. @ 6% each = 90%

1.   Eight less than twice a number is 16. Find the number.

2.   One whole number is two times a second whole number. Together, the two numbers add up to 45. Find both numbers.

3.   One whole number is five times a second whole number. If 7 is added to four times the smaller number, the result is the same as adding one to the larger number. Find both numbers.

4.   Separate 52 into two parts such that twice the smaller is 4 less than the larger.

5.   Haley is four years older than Chris. If Chris's age is represented by $7g$, how would you represent Haley's age?

6.   Maritza is eight years younger than Latrell. Together, their ages add up to 42. How old is Latrell?

7.   Dennis is 35. If Dennis is three years younger than twice Harvey's age, how old is Harvey?

8. Mike is eight years older than Brenda. Jerry is twice Mike's age. If the total of all their ages is 64, how old is Jerry?

9. Four consecutive integers add up to 138. Find them.

10. Determine three consecutive integers such that the sum of four times the first integer plus the second integer equals four times the third.

11. Find three consecutive even integers whose sum is 252.

12. Find three consecutive even integers such that three times the second is 12 more than two times the third.

13. The ages of Kendall, Honey, and LaDainian are in consecutive even-number order. Sixteen less than Honey's age equals 52 less than the sum of Kendall's and LaDainian's ages. Find Kendall's age.

14. The sum of four consecutive odd integers is 80. Find them.

15. Determine three consecutive odd integers such that 19 more than the first is 67 less than three times the last.

16–17. @ 5% each = 10%

16. If $4x + y$ represents the first of three consecutive even integers, find the next two integers.

17. If $4p - 3$ is the first of three consecutive odd integers, find the next two integers.

7
Statistics

In this "information age," we're overwhelmed by facts and figures. The only way to make sense of these large amounts of information is to organize them so that we can make sense of them. That's where statistics comes in. Statistics organizes and interprets the vast mountains of information available to us.

The Greek letter sigma, Σ, is very useful in statistics for it is a shorthand *summation* notation.

Example 1

Find $\displaystyle\sum_{i=1}^{9} i$.

Analysis

The Σ notation tells us to add all the integers, i, from 1 to 9.

Work

$$\sum_{i=1}^{9} i = 1 + 2 + 3 + 4 + 5 + 6 + 7 + 8 + 9 = 45$$

Answer

45

Exercises

1a.   Determine $\displaystyle\sum_{i=1}^{7} i$.   1b.   Find $\displaystyle\sum_{i=4}^{8} i$.

Example 2

Find $\displaystyle\sum_{i=1}^{5} i^2$.

Analysis

Find the sum of the square of i from 1 to 5.

Work

$$\sum_{i=1}^{5} i^2 = 1^2 + 2^2 + 3^2 + 4^2 + 5^2$$
$$= 1 + 4 + 9 + 16 + 25 = 55$$

Answer

55

Exercises

2a.   Find $\displaystyle\sum_{i=1}^{5} i^2$.   2b.   Determine $\displaystyle\sum_{i=2}^{4} i^3$.

Example 3

Find $\displaystyle\sum_{j=2}^{6} 3(j+5)$.

Analysis

Find the sum of the products of 3 and $j+5$ from $j = 2$ to 5.

Work

$$\sum_{j=2}^{5} 3(j+5) = 3(2+5) + 3(3+5) + 3(4+5) + 3(5+5)$$
$$= 21 + 24 + 27 + 30 = 102$$

Answer

102

3a. Determine $\displaystyle\sum_{j=2}^{6} 4(j-1)$.   3b. Find $\displaystyle\sum_{i=4}^{7} 3(i+1)$.

Exercises

Example 4

Find $5\displaystyle\sum_{j=4}^{7} 5(j-3)^2$. and compare the answer

with $5\displaystyle\sum_{j=4}^{7}(j-3)^2$.

Analysis

First, find the sum of the products of 5 and $(j-3)^2$ from $j = 4$ to 7. Then add $(j-3)^2$ from $j = 4$ to 7 and multiply the total by 5.

Work

$$\sum_{j=4}^{7} 5(j-3)^2 = 5(4-3)^2 + 5(5-3)^2 + 5(6-3)^2 + 5(7-3)^2$$
$$= 5(1)^2 + 5(2)^2 + 5(3)^2 + 5(4)^2$$
$$= 5(1) + 5(4) + 5(9) + 5(16)$$
$$= 5 + 20 + 45 + 80 = 150$$

$$5\sum_{j=4}^{7}(j-3)^2 = 5\{(4-3)^2 + (5-3)^2 + (6-3)^2 + (7-3)^2\}$$
$$= 5(1^2 + 2^2 + 3^2 + 4^2)$$
$$= 5(1+4+9+16) = 5(30) = 150$$

Answer

$$\sum_{j=4}^{7}5(j-3)^2 = 5\sum_{j=4}^{7}(j-3)^2 = 150$$

Exercises

4a. Determine $2\sum_{i=2}^{5}(i+4)^2$.   4b. Find $3\sum_{j=6}^{8}(j-3)^3$.

In general, we have just shown that $\sum_{i=1}^{n}ki = k\sum_{i=1}^{n}i$, where k is a constant.

Example 5

The following table shows the number of eggs, x_i, eaten by each member of the Davis family.

Person:	Jamal	Eddie	Morwan	JoAnn	Tanisha
x_i	5	7	4	8	2

Find $\sum_{i=1}^{5}x_i$.

Analysis

In this case, x_i represents the number of eggs eaten by each member of the Davis family. $\sum_{i=1}^{5}x_i$ represents the sum of all the eggs.

Work

$$\sum_{i=1}^{5}x_i = 5 + 7 + 4 + 8 + 2 = 26$$

Answer

26

5a. $x_1 = 4$, $x_2 = 7$, $x_3 = 9$, $x_4 = 3$, $x_5 = 7$. Find $\sum_{i=1}^{5} x_i$.

5b. The following table shows the number of books, x_i, sold by the Elm Street Bookstore during the past week.

Day (i)	1	2	3	4	5	6	7
x_i	124	145	13	243	128	167	212

Find $\sum_{i=1}^{7} x_i$.

Quite often we are interested in a typical product, consumer, voter, and the like. When we talk about "typical," we're looking for something representative of an entire group. We're talking about some sort of central tendency.

We can measure central tendency in three ways:

1. The mean or average
2. The mode
3. The median

The **mean** or **average** is simply the sum of the various pieces of data divided by the number of data. In mathematical notation, the arithmetic mean is indicated by $\bar{x}$, read "x bar."

$$\bar{x} = \frac{\sum_{i=1}^{n} x_i}{n} = \frac{x_1 + x_2 + x_3 + \cdots + x_n}{n}$$

where $n =$ the number of pieces of data.

Example 1

André got 75, 83, 92, 79, 96, and 88 on his exams. To the nearest tenth, what was his test average?

Analysis

Add up all the grades and divide by the number of exams, 6.

Work

$$\overline{x} = \frac{75 + 83 + 92 + 79 + 96 + 88}{6} = \frac{513}{6} = 85.5$$

Answer

85.5

Exercises

1a. During the past eight months, Kim drove the following number of miles per month: 785, 829, 919, 900, 1,032, 820, 764, 904. To the nearest tenth of a mile, what was the mean number of miles she drove per month?

1b. Malika recorded her daily Fahrenheit body temperatures for the past week: 97.6, 98.2, 99.4, 97.2, 99.5, 97.8, 99.1. To the nearest tenth of a degree, what was her mean daily body temperature?

Example 2

The Blue Crockery Company has a quality control program. On a weekly basis, inspectors check damages in their manufactured dishes. During the first 6 weeks of the program, inspectors found the following numbers of dishes damaged, by week:

23, 18, 34, 27, 26, 19

For seven weeks, if management wants to hold the mean number of damaged dishes per week to 24, what is the maximum allowable number of damaged dishes on the seventh week?

Analysis

In this example, we're given the mean, 24, and we want to determine the maximum number of allowable damaged dishes during the seventh week.

Let x = the number of possible dishes damaged the seventh week.

Work

$$24 = \frac{23 + 18 + 34 + 27 + 26 + 19 + x}{7}$$

Multiply by 7: $7\left(24 = \frac{147 + x}{7}\right)$

$$168 = 147 + x$$

Subtract 147: $168 = 147 + x$

Subtract 147: $21 = x$

Symmetric
Property: $x = 21$

Check

$$24 = \frac{23 + 18 + 34 + 27 + 26 + 19 + x}{7}$$

$x = 21$: $24 = \frac{23 + 18 + 34 + 27 + 26 + 19 + 21}{7}$

$$24 = \frac{168}{7}$$

$$24 = 24 \checkmark$$

Answer

21

Exercises

2a.   Rosa is on the school track team and ran the following distances (in miles) during the past six days: 5.6, 4.4, 3.6, 4.2, 5.3, 3.7. If she wants to run at least a daily average of 4.7 miles over a 7-day period, what is the minimum number of miles she has to run on the seventh day?

2b.   The St. Petersburg Pirates scored the following number of points in their last eight basketball games: 54, 82, 71, 66, 52, 85, 58, 63. If they want to have a minimum mean score of 65 over a nine-game series, how many points do they have to score during their ninth game?

Example 3

The following table represents student grades in the final exam (the frequency distribution). Find the mean. Round to the nearest tenth.

In Order, i	Grade, g_i	Number of Students or Frequency, f_i	Weighted Grades, $g_i f_i$
1	94	2	$2 \times 94 = 188$
2	89	4	$4 \times 89 = 356$
3	83	5	$5 \times 83 = 415$
4	81	6	$6 \times 81 = 486$
5	78	2	$2 \times 78 = 156$
6	75	3	$3 \times 75 = 225$
7	72	4	$4 \times 72 = 288$
		$\sum\limits_{i=1}^{7} f_i = 26$	$\sum\limits_{i=1}^{7} g_i f_i = 2{,}114$

Analysis

As you can see, the grades are bunched together. There are two 94s, four 89s, and so on. To find the mean, we'll first multiply each grade, g_i, by the frequency of students obtaining that grade, f_i, to give each grade a weight, $g_i f_i$. Then we'll divide the total of the weighted grades $\sum\limits_{i=1}^{7} g_i f_i$ by the total of the frequencies, $\sum\limits_{i=1}^{7} f_i = 26$

Work

$$\overline{x} = \frac{\sum\limits_{i=1}^{7} g_i f_i}{\sum\limits_{i=1}^{7} f_i} = \frac{2{,}114}{26} = 81.3076 \approx 81.3$$

Answer

81.3

3a. The following table represents the frequency distribution of daily temperatures of Los Alamos, taken at 12 noon. Find the mean temperature. Round to the nearest tenth.

Temperature, t_i (Fahrenheit)	Frequency, f_i	$t_i f_i$
58°	3	
61°	2	
65°	4	
68°	2	
71°	1	
74°	4	
79°	3	
83°	5	

3b. The table below represents the frequency distribution of student weights. Find the mean weight. Round to the nearest tenth.

Pounds, p_i	Frequency, f_i	$p_i f_i$
98	2	
105	1	
109	1	
120	3	
126	2	
132	4	
145	4	
151	3	
158	1	
165	2	
174	2	

Sometimes we want to look at a set of data in a different way. We're not really interested in the mean, or the average. We want to find the middle number. The **median** is the middle number in an ordered set of data.

7.3 The Median

Example 1

In their nine games this season, the Blue Beards high school basketball team scored the following totals per game: 67, 45, 84, 55, 73, 36, 80, 62, 38. Find the median score.

Analysis

Arrange the scores in numerical order. Although scores can be arranged in descending or ascending order, the usual practice is to arrange them in ascending order. Then select the *middle* score.

Work

36, 38, 45, 55, 62, 67, 73, 80, 84

Answer

The median, the middle score, case is 62.

Exercises

1a.   Over the past 15 days, the chickens in coop #8 have produced the following daily numbers of eggs: 18, 23, 16, 26, 28, 22, 17, 29, 19, 13, 25, 30, 18, 15, 19. Find the median.

1b.   The employees in Helmut's office were polled about their sleeping patterns. The 13 responses indicated the number of hours slept the previous night: 7.5, 6, 9, 8.5, 6.5, 9.0, 8, 7.5, 6, 7.5, 5.5, 9, 8.5. Find the median.

It was pretty easy to find the median in Example 1 because we had an odd number of scores. However, what do we do when we have an even number of pieces of data?

Example 2

The Jansen Motor Company has tested the fuel efficiency of its newest car, the Alpha. In 10 tests, the car averaged the following number of miles per gallon:

29.4, 20.1, 18.7, 26.8, 33.8, 20, 25.3, 36.1, 23.5, 32.6

To the nearest tenth, find the median number of miles per gallon.

Analysis

First, arrange the numbers in ascending order. Then, since there are an even number of pieces of data, take the two middle numbers and find their average.

Work

$$18.7, 20, 20.1, 23.5, \underline{\frac{25.3, 26.8}{2}}, 29.4, 32.6, 33.8, 36.1$$

Take the average of the two middle numbers:

$$\frac{25.3 + 26.8}{2} = \frac{52.1}{2} = 26.05 \approx 26.1$$

Answer

26.1

2a.   Students in the physics class received the following grades on their last exam: 80, 75, 65, 92, 56, 79, 48, 58, 92, 85, 76, 68. Find the median.

2b.   The sanitation department has kept a record of the annual number of pounds of garbage generated by families living in several buildings on Stanton Street. The following numbers represent their research: 564, 943, 845, 490, 730, 384, 572, 640, 490, 349. Find the median.

Exercises

The **mode** is the easiest measure in statistics. It's simply the number that occurs most frequently in a given set of data.

7.4 The Mode

Example 1

The recorded Fahrenheit temperatures for the first 15 days of November were: 46, 38, 27, 36, 48, 39, 41, 26, 29, 33, 38, 42, 37, 38, 29. Find the mode.

Analysis

Arrange the temperatures in ascending order. Then, by simple observation, just see which temperatures occur most often.

Work

26, 27, 29, 29, 33, 36, 37, **38**, **38**, **38**, 39, 41, 42, 46, 48

In this example, 38 occurs most often (three times).

Answer

38

Exercises

1–3: Find the mode.

1a. Mr. Jackson, the owner of Jackson's Shoe Store, has recorded the daily sales of shoes in his store for the past 12 days: 35, 19, 23, 32, 19, 28, 35, 31, 23, 18, 17, 19.

1b. The students in Ms. Kazankos' class have received the following grades on their last exam: 88, 76, 62, 99, 74, 66, 85, 65, 77, 85, 98, 100, 63.

Example 2

The Hollywood Burger Palace sold the following number of burgers during the past 10 days: 90, 87, 79, 100, 85, 68, 93, 77, 91, 84. Find the mode.

Analysis

Arrange the numbers in ascending order and see which number occurs most frequently.

Work

68, 77, 79, 84, 85, 87, 90, 91, 93, 100

Answer

No number occurs more frequently than any other number, so, in this case, there is no mode.

Example 3

The Ace Computer Repair Shop repaired the following number of printers per month, during the past 12 months: 18, 54, 23, 32, 15, 18, 37, 32, 29, 27, 37, 33. Find the mode.

Analysis

Arrange the numbers in ascending order and see which number occurs most frequently.

Work

15, **18**, **18**, 23, 27, 29, **32**, **32**, 33, **37**, **37**, 54

Answer

There are three modes: 18, 32, and 37.

Exercises

2/3a.   The Miami Mustard baseball team has scored the following number of runs during the last 12 games: 12, 17, 3, 14, 15, 7, 6, 11, 13, 2, 10, 16.

2/3b.   Dr. Enfume has kept dental records of her patients. She has noted the following numbers of cavities in her patients this past week: 0, 1, 3, 2, 5, 4, 1, 3, 5, 3, 2, 1, 4, 0, 5.

7.5 The Standard Deviation

The mean, median, and mode give us a picture of the typical number in a set of data. This number is representative of the center of the data. However, we often want to see a clearer picture of how the data is spread out—the distribution or dispersion of our data.

For example, let's look at the following temperatures on six days in two cities and compare their means:

	Gotham City	Metropolis
	60°F	50°F
	60°F	70°F
	60°F	75°F
	90°F	80°F
	90°F	85°F
	90°F	90°F
Totals:	450°F	450°F
Mean:	$\dfrac{450}{6} = 75°F$	$\dfrac{450}{6} = 75°F$

The mean temperatures in both cities are the same. However, the temperatures are dispersed differently. In Gotham City, the temperatures are closely grouped around 60°F and 90°F, but the temperatures in Metropolis are spread out.

The *ranges* of temperatures are also different. The range is the difference between the highest and lowest members of the set. The range of temperatures in Gotham City is (90 − 60) = 30°F, while the range in Metropolis is (90 − 50) = 40°F.

Statisticians have developed a mathematical formula—called the **standard deviation**—to help them better analyze the "spread" of the data away from the mean.

Let's go back to the sample temperatures in Gotham City and see how each day's temperature deviates from the mean, 75. After we find the deviation of each day's temperature from the mean, we'll square that deviation.

Example 1

Find the sum of the squares of the deviations from the mean of the temperatures in Gotham City.

Analysis

Use the mean, 75, to find the deviation from the mean and then square that deviation.

Work

Gotham City Temperatures

Temperature of Day, x_i	Mean, $\overline{x}$	Deviation from Mean, $x_i - \overline{x}$	Square of Deviations $(x_i - \overline{x})^2$
60	75	$60 - 75 = -15$	$(-15)^2 = 225$
60	75	$60 - 75 = -15$	$(-15)^2 = 225$
60	75	$60 - 75 = -15$	$(-15)^2 = 225$
90	75	$90 - 75 = +15$	$(+15)^2 = 225$
90	75	$90 - 75 = +15$	$(+15)^2 = 225$
90	75	$90 - 75 = +15$	$(+15)^2 = 225$

$$\sum_{i=1}^{6}(x_i - \overline{x})^2 = 1,350$$

Answer

1,350

1a. Professor Diaz has a first period class and the students have completed a survey indicating their home study time. Find the sum of the squares of the deviations from the mean.

Minutes of Study, x_i	Mean, $\overline{x}$	Deviation from Mean, $x_i - \overline{x}$	Square of Deviations, $(x_i - \overline{x})^2$
90			
65			
52			
104			
78			
47			
38			
62			$\sum_{i=1}^{8}(x_i - x)^2$

1b. Different bakers at the Big Peach Bake Shoppe have baked pies during the past 10 days. Find the sum of the squares of the deviations from the mean.

# of Pies Produced, x_i	Mean, $\overline{x}$	Deviation from Mean, $x_i - \overline{x}$	Square of Deviations, $(x_i - \overline{x})^2$
54			
65			
49			
63			
66			
59			
56			
48			
62			$\displaystyle\sum_{i=1}^{9}(x_i - \overline{x})^2$

If we now take the average of the sum of the squares of the deviations from the mean and then take the square root of that figure, we get the standard deviation, s, or, in Greek, σ (sigma):

$$s = \sqrt{\frac{\displaystyle\sum_{i=1}^{n}(x_i - \overline{x})^2}{n}} = \sqrt{\frac{1,350}{6}} = \sqrt{225} = 15$$

The standard deviation really gives us an idea of the "spread" of the data away from the mean. Now, when we look at the mean as well as the standard deviation of some data, we can get a better idea of the distribution of the data.

Example 2

Find the standard deviation of the temperatures in Metropolis in Example 1. Round to the nearest tenth of a degree. Compare the standard deviation of Metropolis with the standard deviation of Gotham City. According to this comparison, which city has its temperatures more "spread out"?

Analysis

First, find the mean. Then, determine the deviation from the mean for each temperature. Square each deviation. Add up the squares and divide by the number of temperatures, 6. Find the square root of the average and, finally, compare the standard deviations of the two cities.

Work

Metropolis Temperatures

Temperature of Day, x_i	Mean, $\overline{x}$	Deviation from Mean, $x_i - \overline{x}$	Square of Deviations, $(x_i - \overline{x})^2$
50	75	$50 - 75 = -25$	$(-25)^2 = 625$
70	75	$70 - 75 = -5$	$(-5)^2 = 25$
75	75	$75 - 75 = 0$	$(0)^2 = 0$
80	75	$80 - 75 = +5$	$(+5)^2 = 25$
85	75	$85 - 75 = +10$	$(+10)^2 = 100$
90	75	$90 - 75 = +15$	$(+15)^2 = 225$

$$\sum_{i=1}^{6}(x_i - \overline{x})^2 = 600$$

Take the average of the sum of the deviations.

$$\frac{\sum_{i=1}^{6}(x_i - \overline{x})}{n} = \frac{1000}{6} = 166.666$$

Take the square root of the average of the sum of the deviations.

$$s = \sqrt{\frac{\sum_{i=1}^{6}(x_i - \overline{x})^2}{n}} = \sqrt{166.669} \approx 12.9$$

Answer

The standard deviation of the temperatures in Metropolis is 12.9, and the standard deviation of the temperatures in Gotham City is 15. The temperatures in Gotham City are more spread out.

2a. Ms. Clark's students have completed a survey indicating the amount of time they spent watching television yesterday. Find the standard deviation and round to the nearest tenth.

Exercises

Minutes of Study, x_i	Mean, $\bar{x}$	Deviation from Mean, $x_i - \bar{x}$	Square of Deviations, $(x_i - \bar{x})^2$
90			
65			
46			
116			
78			
55			
30			
64			
			$\displaystyle\sum_{i=1}^{8}(x_i - \bar{x})^2$

2b.   The results of a survey of the number of movies some teenagers have seen the past year is indicated in the following table. Find the standard deviation and round to the nearest tenth.

# of Movies, x_i	Mean, $\bar{x}$	Deviation from Mean, $x_i - \bar{x}$	Square of Deviations, $(x_i - \bar{x})^2$
23			
2			
15			
53			
24			
8			
17			
40			
61			
			$\displaystyle\sum_{i=1}^{9}(x_i - \bar{x})^2$

7.6 The Normal Distribution

If we sample natural phenomena—such as weight, intelligence, and height—most of the data are bunched together in the middle and thin out at both the upper and lower ends.

Let's take body weight as one example. Most people weigh close to the average—at the middle. As we move farther and farther away from the middle, there are fewer and fewer heavyweights at the top end and fewer and fewer lightweights at the lower end. The "bell-shaped" curve represents the "normal" range of naturally occurring phenomena.

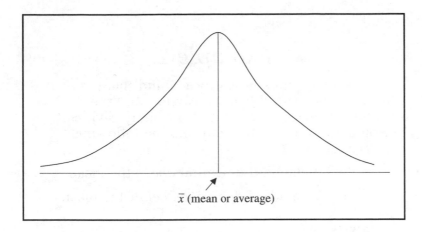

$\overline{x}$ (mean or average)

The following graph represents a normal distribution in which the mean, the median, and the mode are identical. $\overline{x}$ represents the mean, and s represents 1 standard deviation.

1. 68% of the data are within the range of 1 standard deviation from the mean. In other words, 68% of the data occur between $\overline{x} - s$ and $\overline{x} + s$.

2. 95% of the data occur within 2 standard deviations of the mean ($\overline{x} - 2s$ and $\overline{x} + 2s$).

3. 99.5% of the data occur within 3 standard deviations of the mean ($\overline{x} - 3s$ and $\overline{x} + 3s$).

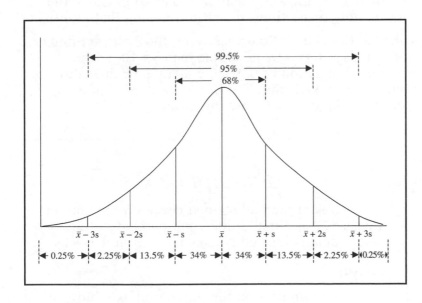

Example 1

In a recent study, it was found that the average 15-year-old had 1.6 cavities and the standard deviation was 0.3. How many cavities would 15-year-olds that fall into the following categories have?

a. 1 standard deviation *above* the mean

b. 2 standard deviations *below* the mean

Analysis

Go back to the mean and add or subtract the requisite number of deviations.

Work

a. 1 standard deviation above the mean:
$\bar{x} + s = 1.6 + 0.3 = 1.9$

b. 2 standard deviations below the mean:
$\bar{x} - 2s = 1.6 - 2(0.3) = 1.6 - 0.6 = 1$

Answers

a. 1.9 cavities

b. 1 cavity

Exercises

1a. It rained last year at an average monthly rate of 9.4 days, and the standard deviation was 0.6. If one particular month was 1 standard deviation above the mean, how many times did it rain during that month?

1b. In the United States last year, the average number of sick days taken by an employee was 13, with a standard deviation of 0.9. How many sick days did employees take if they fell into the category of 2 standard deviations below the mean?

Example 2

Look at the bell-shaped curve on the bottom of page 177 and determine what percent of the sample data is located between $\bar{x} - 2s$ and $\bar{x} + 2s$.

Analysis

The numbers on the horizontal axis indicate the percents between the various standard deviations, so just add up the percents between $\bar{x} - 2s$ and $\bar{x} + 2s$.

Work

Percents between $\bar{x} - 2s$ and $\bar{x} + 2s$:
13.5% + 34% + 34% + 13.5% = 95%.

Answer

95%

2a. What percent of a normal distribution is located between $\bar{x} - s$ and $\bar{x} + 3s$?

2b. Determine what percent of a normal distribution lies between $\bar{x} - 3s$ and $\bar{x} + 2s$?

Exercises

The **percentile** indicates what percent of the total data is located at or below that specific value. The **lower quartile** of a set of data is the 25th percentile. This means that 25% of all the data are located at or below that specific value. The **upper quartile** of a set of data is the 75th percentile, so 75% of all the data are located at or below that specific value.

Percentiles

Example 3

Change $\bar{x} - s$ to a percentile.

Analysis

We want to determine how many data lie below or equal to $\bar{x} - s$, so we simply add up the percents below $\bar{x} - s$.

Work

Percents at or below $\bar{x} - s$:
0.25% + 2.25% + 13.5% = 16%.

Answer

16th percentile

3a. Change $\bar{x} + 2s$ to a percentile.

3b. Change $\bar{x} - s$ to a percentile.

Exercises

Example 4

Weights of 200 people were recorded. If the mean weight was 136 pounds and the standard deviation was 14 pounds, how many people weighed at or below 150 pounds?

Analysis

150 pounds is at $\bar{x} + s$, so we'll change $\bar{x} + s$ to a percentile and then multiply by the number of cases, 200.

Work

Percents below
or equal to $\bar{x} + s$: $0.25\% + 2.25\% + 13.5\%$
                                       $+ 34\% + 34\% = 84\%$

200 cases altogether: $0.84 \times 200 = 168$

Answer

    168

Exercises

4a. A survey of the length of 150,000 telephone calls was made. The average length of call was 14.6 minutes, and the standard deviation was 2 minutes. How many callers spoke less than 10.6 minutes?

4b. 2,300 light bulbs were tested. The mean life of a light bulb was found to be 346 hours with a standard deviation of 40 hours. How many light bulbs burned out in less than 306 hours?

Example 5

In a recent study, the mean price of a quart of milk was $1.43 and the standard deviation for prices of a quart of milk at all local groceries was $0.08. If the Big Bernie Supermarket sells a quart of milk for $1.27, at what percentile is this?

Analysis

$$\$1.43 - \$1.27 = \$0.16$$

If one standard deviation is $0.08, then $0.16 = 2 standard deviations, so all we have to do is to add up the percents up to 2 standard deviations *below the mean.*

Work

$$0.25\% + 2.25\% = 2.5\%$$

Answer

2.5 percentile

5a. The mean life of an automobile tire is 17.3 months, and the standard deviation is 3 months. If a particular tire lasted 11.3 months, what percentile did it fall into?

5b. The average American lives to 76.1 years, with a standard deviation of 5.3 years. If Ms. Chun is 92 years old, in what percentile is she?

Example 6

The Precision Tool Company manufactures widgets. The pieces must be extremely precise. Nevertheless, they vary in length. Any pieces outside of 2 standard deviations of a determined mean length must be discarded. If the company manufactures 22,000 widgets, how many would have to be discarded?

Analysis

We know that 95% of the widgets lie within 2 standard deviations of the predetermined mean length. This means that 5% lie outside of 2 standard deviations, and these pieces must be discarded.

Work

$$0.05 \times 22,000 = 1,100$$

Answer

1,100

6a. The mean weight of an air force cadet should be 165 pounds with a standard deviation of 15 pounds. All candidates outside of 3 standard deviations are rejected. If 43,000 candidates apply, how many would you expect to be rejected?

6b. On an average, people in Oyster Bay City have blood pressure of 135, with a standard deviation of 12. Out of a sample of 9,000 people, how many would you expect to have blood pressure outside of 1 standard deviation from the mean?

Test

1. Determine $\sum_{i=7}^{11} i$.

2. Find $\sum_{i=6}^{9} i^2$.

3. Determine $\sum_{j=2}^{4} 5(j-1)$.

4. Determine $2\sum_{i=3}^{6} (i+4)^2$.

5. The following table shows the number of cars, x_i, sold by Honest Charlie Cars during the past six days:

i	1	2	3	4	5	6
x_i	5	4	3	6	1	7

Find $\sum_{i=1}^{6} x_i$.

6. Janice is on the track team. She recorded the miles she ran each day for the past week: 5.4, 6, 3.7, 6.2, 4.5, 6.1, 3.8. To the nearest tenth of a mile, what was the mean number of miles she ran a day?

7. The Renata Car Company is testing its new model for mileage per gallon. In five tests, the car has been able to operate on the following numbers of miles per gallon: 19, 26, 18, 27, and 22. If the company wants the car to average at least 25 mpg in six tests, what is the minimum miles per gallon on the sixth test?

8. The following table represents the numbers of cancellations per day of air flights during the past month for airlines operating out of Moscow Airport. Find the mean number of daily cancellations.

Cancellations Per Day	# of Days
12	7
15	4
13	1
8	6
17	4
15	5
20	3

9.   NASA launched the following numbers of rockets each year over the past nine years: 9, 4, 7, 3, 10, 5, 4, 3, 6. Find the median number.

10.   During the past 12 years, the Patent Office annually granted the following a number of patents for kitchen gadgets: 345, 123, 99, 118, 201, 137, 67, 89, 88, 60, 180, 213. To the nearest tenth, find the median.

11.   The basketball owners and players in the Central Basketball Association are concerned about the injuries to their players. In a recent study, the following numbers of injuries to players on association teams this past season was noted: 8, 3, 12, 5, 7, 5, 9, 7, 10, 5, 4. Find the mode(s).

12.   During the past 12 months, the Elephantine Press has published the following number of new editions per month: 5, 4, 8, 7, 4, 1, 3, 7, 9, 10, 9, 1. Find the mode(s).

13.   Students in the gym class had to exercise and lift weights. The following table represents the number of pounds lifted by the students in one class. Find the sum of the squares of the deviations from the mean.

# of Pounds Lifted, x_i	Mean, $\overline{x}$	Deviation from Mean, $x_i - \overline{x}$	Square of Deviations, $(x_i - \overline{x})^2$
92			
28			
75			
21			
37			
55			
39			
45			
			$\sum_{i=1}^{8}(x_i - \overline{x})^2$

14. Hector was learning Japanese. He kept a record of the number of new words he learned each day. The following table shows the number of new words he learned during the past nine days. To the nearest tenth, find the standard deviation.

# of Words Learned, x_i	Mean, $\overline{x}$	Deviation from Mean, $x_i - \overline{x}$	Square of Deviations, $(x_i - \overline{x})^2$
11			
9			
8			
13			
6			
17			
8			
16			
11			
			$\sum_{i=1}^{9}(x_i - \overline{x})^2$

15. The Better Driving Association just completed a study of drivers passing through red lights. The association discovered that the average driver went through 5.4 red lights last year and that the standard deviation was 0.8. If a driver was 2 standard deviations below the mean, how many times did he go through a red light?

16. What percent of a normal distribution is located between $\overline{x} - 2s$ and $\overline{x} + s$?

17. Change $\overline{x} + s$ to a percentile.

18. A survey of manufacturing mistakes on automobiles was conducted on 30,000 cars. There was an average of 12 manufacturing errors on each car. The standard deviation was 1.5 errors. How many cars had more than 15 manufacturing errors?

19. The mean life span of a computer is 3.4 years with a standard deviation of 0.6 years. If Mandrake's computer lasted 1.6 years, in what percentile was the life of his computer?

20. The Happy Green Ball Company manufactures tennis balls. If the company rejects all balls outside of 3 standard deviations, how many balls would be rejected out of a total of 40,000?

8

First-Degree Equations

A first-degree equation in two variables—sometimes called a linear equation in two variables—can be represented on a graph as a straight line. Some examples of first degree equations in two variables are

$$y = 3x + 1 \qquad y = x - 2 \qquad x + 3 = -2 \qquad x = 2y - 5$$

$$2x - y = 8$$

In a first-degree equation, the exponent of any variable is 1.

First-degree equations are probably the most common of all equations. When we are given two first-degree equations with two variables and we need to solve for both variables, we need to manipulate both equations simultaneously. There are two basic methods

Definition of a First-Degree Equation in Two Variables

Simultaneous First-Degree Equations

185

of performing these manipulations of simultaneous equations:

1. The method of addition or subtraction

2. The method of substitution

8.1 Number Problems

Some number problems can be solved using only one variable. However, in the following examples, we'll use two variables to get some practice in solving linear equations with two variables.

Example 1

The sum of two numbers is 24. Their difference is 6. Find the numbers.

Analysis

Let's use the **method of addition** to solve this set of simultaneous first-degree equations.

Let x = the larger number.

Let y = the smaller number.

Work

The sum of the numbers is 24:	(1)	$x + y = 24$
The larger number— the smaller number is 6:	+ (2)	$x - y = 6$
Add (1) + (2):		$2x = 30$

Divide by 2:		$x = 15$
Go to (1):	(1)	$x + y = 24$
$x = 15$:	(1)	$15 + y = 24$
Subtract 15:	(1)	$y = 9$

Check

	(1)	$x + y = 24$
$x = 15, y = 9$:	(1)	$15 + 9 = 24$
	(1)	$24 = 24$ ✓
	(2)	$x - y = 6$
$x = 15, y = 9$:	(2)	$15 - 9 = 6$
	(2)	$6 = 6$ ✓

Answer

The larger number is 15; the smaller number is 9.

1a. The sum of two numbers is 12, and their difference is 6. Find both numbers.

1b. The difference of two numbers is 15, and their sum is 29. Find them.

Exercises

Example 2

Three times a larger number added to four times a smaller number is 37. The sum of five times the smaller number and two times the larger number is 34. Find both numbers.

Analysis

Let's use the **method of subtraction** to solve this set of simultaneous first-degree equations.

Let x = the larger number.

Let y = the smaller number.

Work

Three times a larger number added to four times a larger number is 37:	(1)	$3x + 4y = 37$

The sum of five times the
smaller number and two
times the larger number is 34: (2) $5y + 2x = 34$

Here we can't just add or subtract the two
equations and make one of the variables disappear.
First, we must line up the variables and then
multiply each equation by a constant to obtain the
same coefficient in front of the same variable in
both equations.

$$(1) \qquad 3x + 4y = 37$$

Rearrange equation (2): (2) $2x + 5y = 34$

Let's eliminate the x variable.

Multiply (1) by 2: 2 ((1) $3x + 4y = 37$)

Multiply (2) by 3: 3 ((2) $2x + 5y = 34$)

$$(1) \quad 6x + 8y \quad = 74$$

Subtract (2) from (1): $- (2)\oplus \quad 6x \oplus + 15y = \oplus\ 102$

$$-7y = -28$$

Divide by -7: $y = 4$

Go back to original
equation (1): (1) $3x + 4y = 37$

Substitute 4 for y: (1) $3x + 4(4) = 37$

$$(1) \qquad 3x + 16 = 37$$

Subtract 16: (1) $3x = 21$

Divide by 3: (1) $x = 7$

Check

$$(1) \qquad 3x + 4y = 37$$

$x = 7, y = 4$: (1) $3(7) + 4(4) = 37$

$$(1) \qquad 21 + 16 = 37$$

$$(1) \qquad 37 = 37\ \checkmark$$

$$(2) \qquad 5y + 2x = 34$$

$x = 7, y = 4$: (2) $5(4) + 2(7) = 34$

$$(2) \qquad 20 + 14 = 34$$

$$(2) \qquad 34 = 34\ \checkmark$$

Answer

The larger number is 7, the smaller
number 4.

2a. Five times a smaller number added to three times a larger number is 41. The difference between nine times the smaller number and four times the larger number is 8. Find both numbers.

2b. The difference between four times a larger number and five times a smaller number is 13. The sum of six times the larger number and two times the smaller number is 48. Find both numbers.

Example 3

One number is four times another. If the difference between the larger and the smaller is 15, find both numbers.

Analysis

In this case, it's easier to solve directly with one first-degree equation rather than to use two simultaneous equations.

Let x = the smaller number.

Let $4x$ = the larger number.

Work

The difference is 15: $4x - x = 15$

$$3x = 15$$

Divide by 3: $x = 5$

$$4x = 4(5) = 20$$

Check

$$4x - x = 15$$
$$20 - 5 = 15 \checkmark$$

Answer

The larger number is 20, the smaller number is 5.

3a. One number is three times another. If their difference is 30, find both numbers.

3b. One number is five times a second number. Their difference is 28. Find both numbers.

Example 4

The difference between two numbers is 4. Twice the smaller increased by 5 is the same as 4 more than the larger number. Find both numbers.

Analysis

Let's use the **method of substitution** to solve this set of simultaneous first-degree equations.

Let x = the smaller number.

Let y = the larger number.

Work

The difference between two numbers is 4 (larger – smaller):	(1)	$y - x = 4$
Twice the smaller increased by 5 is the same as 4 more than the larger:	(2)	$2x + 5 = y + 4$
Rearrange the terms in (1) by adding x:	(1)	$y = 4 + x$
Substitute $4 + x$ for y in (2):	(2)	$2x + 5 = (4 + x) + 4$
	(2)	$2x + 5 = 8 + x$
Subtract x:	(2)	$x + 5 = 8$
Subtract 5:	(2)	$x = 3$
Go back to original equation (1):	(1)	$y - x = 4$
Substitute 3 for x:	(1)	$y - 3 = 4$
Add 3:	(1)	$y = 7$

Check

	(1)	$y - x = 4$
$x = 3, y = 7$:	(1)	$7 - 3 = 4$
	(1)	$4 = 4$ ✓
	(2)	$2x + 5 = y + 4$
$x = 3, y = 7$:	(2)	$2(3) + 5 = 7 + 4$
	(2)	$6 + 5 = 11$
	(2)	$11 = 11$ ✓

Answer

The larger number is 7; the smaller number is 3.

4a. The sum of two numbers is 15. The sum of two times the smaller number and four times the larger number is 54. Find both numbers.

4b. The sum of two numbers is 13. The difference between five times the larger number and two times the smaller number is 23. Find both numbers.

Example 1

A store sells eight pencils and four erasers for \$6.56. The same store sells five pencils and seven erasers for \$8.42. Determine the cost of one pencil and one eraser.

Analysis

Let p = the price of one pencil.

Let e = the price of one eraser.

Set up two equations.

Work

Eight pencils and four erasers sell for \$6.56:

(1) $8p + 4e = 6.56$

Five pencils and seven erasers sell for \$8.42:

(2) $5p + 7e = 8.42$

Change to cents:

(1) $8p + 4e = 656$

(2) $5p + 7e = 842$

We want to manipulate the two equations to eliminate one of the variables. Let's get rid of the p variable by multiplying equation (1) by 5 and equation (2) by 8.

Multiply (1) by 5:	(1)	$5(8p + 4e = 656)$
Multiply (2) by 8:	(2)	$8(5p + 7e = 842)$

Now we have 40p in both equations.	(1)	$40p + 20e = 3{,}280$
	$-$ (2)	$40p + 56e = 6{,}736$

Subtract (2) from (1):		$-36e = -3{,}456$
Divide by -36:		$e = 96$ or, in decimals, $\$0.96$

Now that we know that $e = \$0.96$, we'll substitute for e in one of our original equations.

Let's arbitrarily take (1):	(1)	$8p + 4e = 6.56$
Let $e = 0.96$:	(1)	$8p + 4(0.96) = 6.56$
	(1)	$8p + 3.84 = 6.56$
Subtract 3.84:	(1)	$8p = 2.72$
Divide 8:	(1)	$p = 0.34$

Check

	(1)	$8p + 4e = 6.56$
$p = 0.34, e = 0.96$:	(1)	$8(0.34) + 4(0.96) = 6.56$
	(1)	$2.72 + 3.84 = 6.56$
	(1)	$6.56 = 6.56$ ✓
	(2)	$5p + 7e = 8.42$
$p = 0.34, e = 0.96$:	(2)	$5(0.34) + 7(0.96) = 8.42$
	(2)	$1.70 + 6.72 = 8.42$
	(2)	$8.42 = 8.42$ ✓

Answer

1 pencil costs $0.34, 1 eraser costs $0.96

Exercises

1a. Six apples and seven peaches cost $5.80. At the same time, eight apples and five peaches cost $5.48. How much is each apple and each peach?

1b. A fruit store has a sale. A customer can buy five bananas and nine cantaloupes for $12.80 or eight bananas and four cantaloupes for $6.96. Find the cost of one banana and one cantaloupe.

Example 2

The Honest Charlie Auto Dealership sells used automobiles. They sell only two types of cars—an American model and an import—and they sell 500 cars per year. They make a profit of $550 on each American car and a profit of $420 on each import for a total annual profit of $251,600. How many cars of each type do they sell?

Analysis

Honest Charlie sells a total of 500 cars per year.

Let x = the number of American cars.

Let $500 - x$ = the number of imports.

Work

Type of Car	Number Sold	Profit Per Car	Total Profit
American	x	550	$550x$
Import	$500 - x$	420	$420(500 - x)$

Total profit is
$251,600:

$$550x + 420(500 - x) = 251,600$$
$$550x + 210,000 - 420x = 251,600$$
$$130x + 210,000 = 251,600$$

Subtract 210,000:
$$130x = 41,600$$

Divide by 130:
$$x = 320$$
$$500 - x = 180$$

Check

$$550x + 420(500 - x) = 251,600$$

$x = 320$,
$500 - x = 180$:
$$550(320) + 420(180) = 251,600$$
$$176,000 + 75,600 = 251,600$$
$$251,600 = 251,600 \checkmark$$

Answer

320 American cars, 180 imports

2a. The Big Brother Battery Company manufactures automobile batteries. The company manufactures two types of batteries, type A and type B. It manufactures 6,000 batteries a month and makes a total monthly profit of $52,000. Its profit margin on each type A battery is $12, and its profit margin on each type B battery is $2. How many batteries of each type does it manufacture?

2b. The Omaha Corporation produces 4,500 computer printers per month. It manufactures the Perfect Printer and the Perfect Printer +. The company makes a monthly profit of $65,000 on sales of its two printers. If its profit margin is $13 on each Perfect Printer and $18 on each Perfect Printer +, how many printers of each type are sold?

8.3 Upstream-Downstream Problems

Example 1

An airplane can fly at the rate of 800 mph in still air. If it flies with the wind, it can cover 6,560 miles in the same time that it can fly 6,540 miles against the wind. What is the wind speed?

Analysis

Let w = the rate of the wind (in mph).

Let t = time (in hours).

Take note of the following:

1. Distance = Rate × Time.

2. If the plane flies with the wind, add the two speeds together to get the resultant speed.

3. If the plane flies against the wind, subtract the wind speed from the plane's speed to obtain the resultant speed.

	Plane's Rate in Still Air, r	Rate of Wind, w	Resultant Rate, $r \pm w$	Time, t	Distance, $(r \pm w)t$
With wind	800	w	$800 + w$	t	$(800 + w)t$
Against wind	800	w	$800 - w$	t	$(800 - w)t$

Work

If the plane flies with the wind, it covers 6,560 miles:

(1) $(800 + w)t = 6{,}560$

If the plane flies against the wind, it covers 6,240 miles:

(2) $(800 - w)t = 6{,}240$

To eliminate wt, add the two equations:

(1) $800t + wt = 6{,}560$

$+$ (2) $\underline{800t - wt = 6{,}240}$

$1{,}600t = 12{,}800$

Divide by 1,600: $t = 8$

Go back to (1):

(1) $(800 + w)t = 6{,}560$

(1) $800t + wt = 6{,}560$

Substitute 8 for t:

(1) $800(8) + w(8) = 6{,}560$

(1) $6{,}400 + 8w = 6{,}560$

Subtract 6,400:

(1) $8w = 160$

Divide by 8:

(1) $w = 20$

Check

(1) $(800 + w)t = 6{,}560$

$w = 20,\ t = 8$:

(1) $(800 + 20)8 = 6{,}560$

(1) $(820)8 = 6{,}560$

(1) $6{,}560 = 6{,}560$ ✓

(2) $(800 - w)t = 6{,}240$

$w = 20,\ t = 8$:

(2) $(800 - 20)8 = 6{,}240$

(2) $(780)8 = 6{,}240$

(2) $6{,}240 = 6{,}240$ ✓

Answer

The wind speed is 20 mph.

1a. An airplane can fly at the rate of 380 mph in still air. If it flies against the wind, it can fly 1,800 miles in the same time that it can fly 2,000 miles with the wind. Find the speed of the wind.

1b. A 747 airliner can cruise at 750 mph in still air. If it can fly 4,620 miles with a tailwind in the same time that it can fly 4,380 miles against the wind, determine the speed of the wind and the time for the trip.

1c. A boat that can travel 6 mph in still water moves a distance of 36 miles upstream. If it can cover 108 miles downstream in the same time, find the rate of the current and the time for each trip.

1d. The *Pequod*, a ship, can move at the rate of 27 mph in still water. If it can travel 75 miles against the current in the same time that it can travel 87 miles with the current, find the time for each trip and the rate of the water current.

Example 2

Maria travels upstream by boat a distance of 60 miles in 15 hours. The return trip takes 12 hours. Find the boat's rate in still water and the rate of the water current.

Analysis

Let r = the rate of the boat in still water (mph).

Let c = the rate of the water current (mph).

Let the time upstream = 15 (hours).

Let the time downstream = 12 (hours).

"Upstream" means travel against the current, so we must subtract the rate of the water current from the boat's speed. "Downstream" means travel with the current, so we must add the boat's speed to the water current.

	Boat's Rate in Still Water, r	Rate of Current, c	Resultant Rate, $r \pm w$	Time, t	Distance, $(r \pm w)t$
Upstream	r	c	$(r - c)$	15	$(r - c)15$
Downstream	r	c	$(r + c)$	12	$(r + c)12$

Work

Maria travels upstream
by boat a distance of
60 miles in 15 hours: (1) $(r - c)15 = 60$

The downstream trip
takes 12 hours: (2) $(r + c)12 = 60$

           (1) $15r - 15c = 60$

           (2) $12r + 12c = 60$

     Let's try to eliminate one of the variables. If we multiply (1) by 4 and (2) by 5, we obtain $60r$ in both equations, and we can then eliminate the r variable.

Multiply (1) by 4:   (1) $4(15r - 15c = 60)$

Multiply (2) by 5:   (2) $5(12r + 12c = 60)$

Now that we have
$60r$ in both equations,   (1) $60r - 60c = 240$
we can add to $+$ (2) $60r + 60c = 300$
eliminate c: $120r \qquad\qquad = 540$

Divide by 120: $r = 4.5$

Go back to original
equation (1) and   (1) $(r - c)15 = 60$
substitute 4.5 for r: (1) $(4.5 - c)15 = 60$

           (1) $67.5 - 15c = 60$

Subtract 67.5:   (1) $-15c = -7.5$

Divide by −15:   (1) $c = 0.5$

Check

           (1) $(r - c)15 = 60$

$r = 4.5, c = .5$:   (1) $(4.5 - 0.5)15 = 60$

           (1) $(4)15 = 60$

           (1) $60 = 60$ ✓

           (2) $(r + c)12 = 60$

$r = 4.5, c = .5$:   (2) $(4.5 + 0.5)12 = 60$

           (2) $(5)12 = 60$

           (2) $60 = 60$ ✓

Answer

     The boat travels at the rate of 4.5 mph in still water, and the current moves at the rate of 0.5 mph.

2a. A new airplane can fly 2,620 miles against the wind in 4 hours. If it can fly 2,740 miles with the wind in the same time, determine the wind speed and the airplane's speed in still air.

2b. A boat can travel 60 miles upstream in 6 hours. If it can move the same distance with the current in 4 hours, find the rate of the current and the speed of the boat in still water.

2c. The Army's new transport plane can fly 3,600 miles against the wind in 7.2 hours. If it flies with the wind, it can cover the same distance in 6 hours. What is the rate of the plane in still air and the speed of the wind?

2d. The *Marlin*, a fishing vessel, can travel 40 miles against a water current in 5 hours. If it takes the boat 4 hours to travel 48 miles with the water current, find the boat's rate in still water.

8.4 Digit Problems

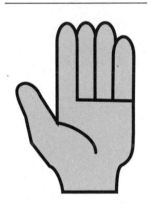

The ten digits we use—0, 1, 2, 3, 4, 5, 6, 7, 8, 9— are undoubtedly derived from our ten counting fingers. The number 74 represents 7 tens and 4 units:

$$74 = 7(10) + 4$$

tens' digit ⎯⎯⎯⎯⎯⎯ units' digit

Similarly, the number 283 represents 2 hundreds, 8 tens, and 3 units:

$$283 = 2(100) + 8(10) + 3(1)$$

hundreds'　　tens'　　units'
digit　　　　digit　　digit

If we want to represent an unknown two-digit number, let's represent the tens' digit by t and the units' digit by u.

$$10t + u$$

10 times the tens' digit + the units' digit

However, let's not make the mistake of writing the two-digit number as tu because tu means the product of t and u, and we don't want that.

From now on, let's represent an unknown two-digit number as $10t + u$. For example, $t = 5$ and $u = 2$, then $10t + u = 10(5) + 2 = 52$.

The sum of the digits, however, is not the same thing as the two-digit number itself. The sum of the digits of the generalized two-digit number $10t + u$ is $t + u$. In the case when $t = 5$ and $u = 2$, the sum of the digits, $t + u$, equals $5 + 2 = 7$.

In a two-digit number, if the tens' digit is larger than the units' digit, then the **difference of the digits** is $t - u$. If the two-digit number is 52, then the difference of the digits is 3.

If the original two-digit number is $10t + u$, then, if we **reverse the digits**, the new number formed is $10u + t$. If the original two-digit number is 52, then the new number formed by reversing the digits is 25.

Example 1

In a two-digit number, the sum of the digits is 14 and the difference of the digits is 2. Find the number if the tens' digit is larger than the units' digit.

Analysis

Let t = the tens' digit and let u = the units' digit.

Work

The sum of the digits is 14:   (1) $t + u = 14$

The difference is 2 (Remember that the tens' digit is larger so the difference is a positive 2.):   (2) $t - u = 2$

                              (1) $t + u = 14$

Let's add the two equations so as to eliminate u: (2) $\underline{t - u = 2}$

                                          $2t = 16$

Divide by 2: $t = 8$

Substitute 8 for t in (1):   (1) $t + u = 14$

                              (1) $8 + u = 14$

Subtract 8:   (1) $u = 6$

Check

                              (1) $t + u = 14$

$t = 8$, $u = 6$:   (1) $8 + 6 = 14$

                              (1) $14 = 14$ ✓

	(2) $t - u = 2$
$t = 8$, $u = 6$:	(2) $8 - 6 = 2$
	(2) $2 = 2$ ✓

Answer

   86

Find the two-digit number described.

1a. The sum of the digits is ten and the difference is six. The tens' digit is larger than the units' digit.

1b. The tens' digit is larger than the units' digit. The sum of the digits is 13, and the difference is 1.

Example 2

In a two-digit number, five times the tens' digit added to four times the units' digit is 57. The difference between eight times the units' digit and two times the tens' digit is 6. Find the original two-digit number.

Analysis

   Let t = the tens' digit.

   Let u = the units' digit.

Work

Five times the tens' digit added to four times the units' digit is 57:	(1)	$5t + 4u = 57$
The difference between eight times the units digit and two times the tens' digit is 6:	(2)	$8u - 2t = 6$
Rearrange (2):	(2)	$-2t + 8u = 6$

Let's eliminate u, so we have to get the same coefficients.

Multiply (1) by 2:	(1)	$2(5t + 4u = 57)$
Multiply (2) by 5:	(2)	$5(-2t + 8u = 6)$
	(1)	$10t + 8u = 114$
	+ (2)	$-10t + 40u = 30$
Add (1) and (2):		$48u = 144$
Divide by 48:		$u = 3$

	(1)	$5t + 4u = 57$
Let $u = 3$:	(1)	$5t + 4(3) = 57$
	(1)	$5t + 12 = 57$
Subtract 12:	(1)	$5t = 45$
Divide by 5:	(1)	$t = 9$

Check

	(1)	$5t + 4u = 57$
$t = 9, u = 3$:	(1)	$5(9) + 4(3) = 57$
	(1)	$45 + 12 = 57$
	(1)	$57 = 57$ ✓
	(2)	$8u - 2t = 6$
$t = 9, u = 3$:	(2)	$8(3) - 2(9) = 6$
	(2)	$24 - 18 = 6$
	(2)	$6 = 6$ ✓

Answer

93

Find the two-digit number described.

Exercises

2a. Five times the tens' digit plus six times the units' digit is 38. Two times the tens' digit added to seven times the units' digit is 29.

2b. Four times the tens' digit plus three times the units' digit is 34. The difference between five times the tens' digit and two times the units' digit is 31.

Example 3

In a two-digit number, the units' digit is four times the tens' digit. If the digits are reversed, the new number is 54 more than the original number. Find the original two-digit number.

Analysis

Let t = the tens' digit.

Let u = the units' digit.

The original two-digit number is $10t + u$. If the digits are reversed, u = the tens' digit and t = the units' digit, so the new two-digit number is $10u + t$.

Work

The units' digit is four times the tens' digit:

(1) $u = 4t$

If the digits are reversed, the new number is 54 more than the original number:

(2) $10u + t = 10t + u + 54$

The easiest way to solve these first degree equations is to substitute $4t$ for u in (2):

(2) $10(4t) + t = 10t + (4t) + 54$

(2) $40t + t = 10t + 4t + 54$

(2) $41t = 14t + 54$

Subtract $14t$:

(2) $27t = 54$

Divide by 27:

(2) $t = 2$

Substitute 2 for t in (1):

(1) $u = 4t$

(1) $u = 4(2)$

(1) $u = 8$

Check

(1) $u = 4t$

$t = 2, u = 8$:   (1) $8 = 4(2)$

(1) $8 = 8$ ✓

(2) $10u + t = 10t + u + 54$

$t = 2, u = 8$:   (2) $10(8) + 2 = 10(2) + 8 + 54$

(2) $80 + 2 = 20 + 8 + 54$

(2) $82 = 82$ ✓

Answer

28

Exercises

Find the two-digit number described.

3a.   The units' digit is four times the tens' digit. If the digits are reversed, the new number is 54 more than the original number.

3b.   The units' digit is six less than the tens' digit. If the digits are reversed, the new number is 147 less than two times the original number.

Example 1

Ms. Jackson invested a certain amount of money in ABC Corporation bonds paying simple annual interest at 8%. She placed $2,000 more than this amount in XYZ Corporation bonds paying simple annual interest at 6%. If her total annual income from these two sources was $820, how much did she invest in each corporation?

Analysis

Let x = the principal invested in ABC Corporation.

Let $x + 2,000$ = the principal invested in XYZ Corporation.

Income = Principal × Rate × Time = prt

Investment	Principal, p	Annual Rate, r	Time, t (years)	Income = prt
ABC	x	8%	1	8%$(x)(1)$
XYZ	$x + 2,000$	6%	1	6%$(x + 2,000)(1)$

Work

Her total annual income from these two sources was $820:
$$8\%(x)(1) + 6\%(x + 2,000)(1) = 820$$

Change to decimals:
$$0.08x + 0.06(x + 2,000) = 820$$

Multiply by 100:
$$8x + 6(x + 2,000) = 82,000$$

$$8x + 6x + 12,000 = 82,000$$

$$14x + 12,000 = 82,000$$

Subtract 12,000:	$14x = 70,000$
Divide by 14:	$x = 5,000$
	$x + 2,000 = 7,000$

Check

$$8\%(x) + 6\%(x + 2,000) = 820$$

$x = 5,000$: $\quad 0.08(5,000) + 0.06(7,000) = 820$

$$400 + 420 = 820$$

$$820 = 820 \checkmark$$

Answer

   Ms. Jackson invested $5,000 in ABC corporate bonds and $7,000 in XYZ corporate bonds.

Exercises

1a.   Holly Kim invested some money in real estate at an annual return of 9% and twice this amount in a bank at an annual rate of 6%. If her total annual return is $1,260, determine both investments.

1b.   Jessica Maynard invested some money at an annual rate of 7% and $3,000 more than this amount invested at an annual rate of 6%. Find both amounts if the total annual interest is $700.

1c.   Carlos Pereira has some money in municipal bonds at an annual rate of 6.5% and $4,000 more than twice this amount in U.S. government bonds at an annual rate of 7%. If his total annual interest is $1,100, how much did he invest in each place?

1d.   Julius Sharpe placed some money in a mutual fund at an annual rate of 7% and $5,000 less than three times this amount in state bonds at an annual rate of 8%. If his total annual interest from these investments is $2,080, find the amount invested at each rate.

Example 2

   Mickey Fried has deposited a total of $9,000 in special and regular savings accounts. His special savings account offers an 8.5% annual return, and the regular savings account offers a 5% annual return. If the special account yields an annual return that is $117 more return than the 5% regular savings account annual return, find the amounts he invested in both places.

Analysis

The total amount of money invested was $9,000.

Let x = the amount of money invested in a special savings account.

Let $9,000 - x$ = the amount he invested in a regular savings account.

The income (return) equals (principal) × (rate) × (time) = prt.

Investment	Principal, p	Annual Rate, r	Time, t (years)	Income = prt
Special account	x	8.5%	1	8.5%(x)(1)
Regular account	$9,000 - x$	5%	1	5%($9,000 - x$)(1)

Work

The 8.5% special savings account yields an income that is $117 greater than his income from the 5% regular savings account:

$$8.5\%(x)(1) = 5\%(9,000 - x)(1) + 117$$

Change % to decimals:

$$0.085x = 0.05(9,000 - x) + 117$$

Multiply by 1,000:

$$85x = 50(9,000 - x) + 117,000$$

$$85x = 450,000 - 50x + 117,000$$

$$85x = 567,000 - 50x$$

Add 50x:

$$135x = 567,000$$

Divide by 135:

$$x = 4,200$$

$$9,000 - x = 4,800$$

Check

$$8.5\%(x) = 5\%(9,000 - x)$$

$x = 4,800$, $9,000 - x = 4,200$:

$$0.085(4,200) = 0.05(4,800) + 117$$

$$357 = 240 + 117$$

$$357 = 357 \checkmark$$

Answer

$4,200 at 8.5%, $4,800 at 5%

2a. The Circle Investment Corporation has $250,000 to deposit. It deposited some money at an annual rate of 8% and the remainder at an annual rate of 9%. The annual income derived from the 8% account is $6,400 more than the annual income derived from the 9% account. Determine the amounts of money invested at each rate.

2b. The trustees of McKinley College have $500,000 to invest. Some money is to be invested at an annual rate of 8% and the remainder at an annual rate of 7.5%. If the annual income derived from the 7.5% account is $14,250 more than the annual income derived from the 8% account, find the monies to be invested at each rate.

2c. Darryl Goodin has $18,000 to invest. He wants to invest some money at an annual rate of 9% and the remainder at an annual rate of 7.25%. The annual income from the 7.25% account is $482.50 less than the annual income derived from the 9% account. Find the investments in each place.

2d. Isabella Rossi placed some of her $15,000 in a savings bank at an annual rate of 8.25% and the remainder in a private business venture at an annual rate of 11%. If her annual income from the bank investment is $495 less than the annual income from the private investment, how much did she invest in each place?

Test

1. The sum of two numbers is 32 and their difference is 14. Find the numbers.

2. Three times a larger number added to four times a smaller number equals 44. The difference between two times the smaller number and the larger number is two. Find the two numbers.

3. Four times a larger number added to twice a smaller number equals 58. The difference between nine times the smaller number and three times the larger number is 30. Find the two numbers.

4. One number is four times another number. The difference between three times the larger number and two times the smaller number is 20. Find the two numbers.

5. The Malibu Department Store sells eight ties and seven shirts for $201. At the same time, the store sells nine ties and six shirts for $183. Find the cost of one shirt and one tie.

6. Helene's sells four blouses and five skirts for $262. At the same time, the store sells seven blouses and four

skirts for $278. Find the cost of one blouse and one skirt.

7. The Village Book Store sells a total of 400 books a week. It makes a profit of $5 per each softcover book and $6 per each hardcover book. The weekly profit is $2,080. How many softcover and hardcover books did the store sell?

8. The Ben Franklin Eyeglass Factory sells wire-frame eyeglasses and plastic-frame eyeglasses. They sell a weekly total of 250 eyeglasses for a weekly profit of $2,880. If they make a profit of $12 on each wire-frame pair and a $9 profit on each plastic-frame pair, how many eyeglasses of each type did they sell?

9. A sailboat can travel at the rate of 40mph in still water. Together with a current, it can move 368 miles in the same time that it can move 272 miles against the current. Find the current speed.

10. An airplane flies 4,340 miles against the wind in seven hours and 4,080 miles in six hours with the wind. Find the speed of the airplane.

11. A Mississippi steamboat travels downstream a distance of 240 miles in eight hours. If it can travel the same distance upstream in 10 hours, find the rate of the current and the rate of the boat in still water.

12. The *Whale*, a tugboat, can move 40 miles in 10 hours against the current. If it can cover the same number of miles downstream in 4 hours, find the rate of the current.

13. In a two-digit number, the sum of the digits is 12 and the difference is six. If the tens' digit is larger, what is the two-digit number?

14. In a two-digit number, three times the tens' digit added to five times the units digit equals 53. The difference between four times the tens' digit and two times the units' digit is 10. What is the two-digit number?

15. In a two-digit number, the difference between eight times the tens' digit and three times the units' digit is 11. At the same time, the difference between five times the units' digit and two times the tens' digit is 27. Find the two-digit number.

16. In a two-digit number, the sum of the digits is 10. If the digits are reversed, the new number is 18 less than the original number. What was the original number?

17. Jester invests some money at an annual rate of 7% and $4,000 more than this amount at an annual rate of 6%. If the total annual return is $890, how much did he invest at each rate?

18. Manfredo invests some money in real estate at an annual rate of 11% and $5,000 more than twice this amount in U.S. Treasury Bills at an annual rate of 7%. If his total annual return is $4,100, how much does he invest in each place?

19. Jaqueta has $10,000 to invest. She puts some money in a mutual fund at an annual rate of 10% and the remainder into common stocks paying an annual return of 7%. If her total annual income is $910, how much does she invest in each place?

20. The X Corporation has $50,000 to invest. It places some money into Turkish bonds at an annual rate of 12% and the remainder into British bonds paying an annual interest rate of 9%. If the annual return on the British bonds was $2,610 more than the annual return on the Turkish bonds, how much was invested in each place?

9

Ratios, Proportions, and Variations

A **ratio** is simply a comparison of two numbers. For example, let's say that in the general population the ratio of left-handed people to right-handed people is 3:16. We can look at this statistic in several ways. First, we can conclude that there are 3 left-handed people for every 16 right-handed people. We can also conclude

that for every 19 people (3 + 16), 3 are left-handed and 16 are right-handed.

A ratio may also be expressed as a fraction:

$$3:16 = \frac{3}{16}$$

We can reduce ratios to *simplest* form when both terms of the ratio are whole numbers and there is no common multiple between them:

$$6:42 = \frac{6}{42} = \frac{1}{7} = 1:7$$

Example 1

Simplify $28x^2:7x$.

Analysis

Rewrite the ratio as a fraction and then reduce.

Work

$$28x^2:7x = \frac{28x^2}{7x} = \frac{4x}{1} = 4x:1$$

Answer

$4x:1$

Exercises

Reduce the ratios to the simplest terms.

1a. $63y^5:49y^2$ 1b. $48a^7:32a^3$

Example 2

Simplify $24:60$.

Analysis

Rewrite the ratio as a fraction and then reduce.

Work

$$24:60 = \frac{24}{60} = \frac{2}{5} \text{ or } 2:5$$

Answer

$$2:5 \text{ or } \frac{2}{5}$$

Reduce the ratios to simplest terms.

2a.   5.6 : 4.8   2b.   2.4 : 1.6

Example 3

Simplify $\dfrac{4}{9} : \dfrac{3}{5}$.

Work

$$\frac{4}{9} \div \frac{3}{5} = \frac{4}{9} \times \frac{5}{3} = \frac{20}{27} = 20 : 27$$

Answer

$$20 : 27 \text{ or } \frac{20}{27}$$

Reduce the ratios to simplest terms.

3a.   5/6 : 2/3   3b.   4/9 : 3/5

Let's say we want to compare 7 inches with 2 centimeters. Before we can compare the two, we have to change the measures to a common unit. One inch = 2.54 centimeters, so 7 inches = (7)(2.54) = 17.78 cm.

$$\frac{7 \,\text{in.}}{2 \,\text{cm}} = \frac{(7)(2.54)\,\text{cm}}{2 \,\text{cm}} = \frac{17.78}{2} = \frac{8.89}{1}$$

In most cases, we usually change the larger unit into smaller units first and then compare the two measures.

The following table summarizes some of the most important units of measurement.

English Units	Metric Units
Distance	
1 foot = 12 inches	1 meter = 100 centimeters
1 yard = 3 feet	1 meter = 1,000 millimeters
1 yard = 36 inches	1 kilometer = 1,000 meters
Mass	
1 pound = 16 ounces	1 kilogram = 1,000 grams
1 ton = 2,000 pounds	
Volume	
1 quart = 2 pints	1 liter = 1,000 milliliters
1 gallon = 4 quarts	

Example 4

Simplify 4 feet : 6 yards.

Analysis

Change 6 yards to feet so that both measures are in feet (1 yard = 3 feet).

Work

$$4 \text{ feet} : 6 \text{ yards} = \frac{4 \text{ feet}}{6 \text{ yards}} = \frac{4 \text{ feet}}{6 \times 3 \text{ feet}} = \frac{4 \text{ feet}}{18 \text{ feet}} = \frac{2}{9}$$

or 2 : 9

Answer

$\frac{2}{9}$ or 2 : 9

Exercises

Reduce the ratios to simplest terms.

4a. 3 yards : 5 feet   4b. 48 inches : 2 yards

Example 5

Simplify 250 centimeters : 6 meters.

Analysis

Change 6 meters to centimeters (1 meter = 100 centimeters).

Work

$$250 \text{ centimeters} : 6 \text{ meters} = \frac{250 \text{ centimeters}}{6 \text{ meters}}$$

$$= \frac{250 \text{cm}}{6 \times 100 \text{cm}}$$

$$= \frac{250 \text{cm}}{600 \text{cm}} = \frac{5}{12} \text{ or } 5 : 12$$

Answer

$\frac{5}{12}$ or 5 : 12

Exercises

Reduce the ratios to the simplest terms.

5a. 500 centimeters : 4 meters

5b. 4 kilometers : 5,000 meters

Example 6

Simplify 6 gallons : 3 quarts

Analysis

Change gallons to quarts (1 gallon = 4 quarts).

Work

$$6 \text{ gallons} : 3 \text{ quarts} = \frac{6 \text{ gallons}}{3 \text{ quarts}} = \frac{6 \times 4 \text{ quarts}}{3 \text{ quarts}}$$

$$= \frac{24 \text{ quarts}}{8 \text{ quarts}} = \frac{3}{1} \text{ or } 3 : 1$$

Answer

$\frac{3}{1}$ or 3 : 1

Reduce the ratios to the simplest terms.

6a. 8 pints : 2 quarts   6b. 6,000 milliliters : 2 liters

Exercises

Thus far, our problems were fairly straight-forward. However, there are other ratio problems which take some thought.

Example 7

Thirty math books were distributed to an algebra class at the beginning of the semester. Four books were lost and the rest were returned.

a. In simplest terms, what is the ratio of the number of lost books to the number of books distributed?

b. In simplest terms, what is the ratio of returned books to lost books?

Analysis

Number of distributed books = 30.

Number of lost books = 4.

Number of returned books = 26.

Work

a. $\dfrac{\text{Lost books}}{\text{Distributed books}} = \dfrac{4}{30} = \dfrac{2}{15}$ or $2:15$

b. $\dfrac{\text{Returned books}}{\text{Lost books}} = \dfrac{26}{4} = \dfrac{13}{2}$ or $13:2$

Answers

a. $\dfrac{2}{15}$ or $2:15$ b. $\dfrac{13}{2}$ or $13:2$

Exercises

7a. Out of 340 people in an audience, 70 were *left*-handed. In simplest terms, what was the ratio of *right*-handed people to the total audience?

7b. In a survey, out of 900 people, 60 were smokers. In simplest terms, what was the ratio of smokers to nonsmokers?

Example 8

Montel and Carlos own a gasoline station. They agree to split the profits in the ratio $5:3$, with Montel getting the greater share. In one week, if they make a profit of $2,568, how much does each man receive?

Analysis

Let $5x$ = Montel's share.

Let $3x$ = Carlos' share.

Work

Together, they make a profit of $2,568:

$$5x + 3x = 2{,}568$$

$$8x = 2{,}568$$

Divide by 8: $x = 321$

$5x = 1{,}605$ (Montel's share)

$3x = 963$ (Carlos' share)

Check

$$5x + 3x = 2{,}568$$
$$1{,}605 + 963 = 2{,}568$$
$$2{,}568 = 2{,}568 \checkmark$$

Answer

Montel makes \$1,605; Carlos makes \$963.

Exercises

8a. Shawn and Gisela are partners in a software company, and they agree to divide their income in the ratio of 6:7, with Gisela getting the greater share. If they made \$156,000 last year, what was Shawn's share?

8b. The state and the city divide the sales tax in the ratio of 5:2, with the state getting the larger share. If \$3,150,000,000 in sales taxes was collected last year, what was the state's share?

Example 9

If the Union City Little League baseball team had a winning games:total games ratio of 3:5 and they lost eight games, what was the total number of games played during the season?

Analysis

Let x = the common ratio.

$$\frac{3x}{5x} = \frac{\text{Number of games won}}{\text{Total number of games played}}.$$

Then $\dfrac{2x}{5x} = \dfrac{\text{Number of games lost}}{\text{Total number of games played}}.$

Work

The team lost eight games: $2x = 8$

Divide by 2: $x = 4$

$5x = 20$ (total number of games played)

Answer

20

9a.   NASA has had a successful launch : failure to launch ratio of 15 : 2. If they succeeded in launching 75 space missions, what was the total number of space mission attempts?

9b.   Three out of 19 students at Gramercy College are chemistry majors. If 1,280 students are not chemistry majors, how many students are there at Gramercy College?

Example 10

The ratio of teachers to students at Central High is 1 : 20. If 500 new students are admitted, there will be 24 times as many new students as teachers. How many students and teachers are there in the school now?

Analysis

The number of teachers will remain constant.

Let x = the number of teachers now and in the future.

Let $20x$ = the number of students now.

Let $20x + 500$ = the number of students in the future.

Work

The number of teachers, x, remains constant.

In the future there will be 24 times as many students a teachers:

$$20x + 500 = 24\,(x)$$

Subtract $20x$: $500 = 4x$

Divide by 4: $x = 125$ (teachers now and in the future)

$20x = 2,500$ (students now)

Answer

125 teachers, 2,500 students

10a. A recipe calls for 6 cups of flour for every egg. If 10 cups of flour are added, there will be 8 times as many new cups of flour as the current number of eggs. How many eggs are there in the mix at present? Hint: The number of eggs remains the same.

10b. The ratio of managers to assembly line workers in a factory is $1:5$. If 15 new workers are added, there will be 6 times as many new assembly line workers as the number of current managers. How many workers are currently employed?

If two ratios are equal, we have a **proportion**. For example, if we simplify the ratio 18 oranges: 32 oranges, we get the proportion

$$\frac{18}{32} = \frac{9}{16}$$

In general terms,

$$\frac{a}{b} = \frac{c}{d} \text{ is the same as } a:b=c:d$$

a, b, c, and d are, respectively, called the first, second, third, and fourth terms of the proportion. The two outer terms are called the **extremes**, and the two inner terms are called the **means**:

Means
$$a:b=c:d$$
Extremes

Do these ratios form proportions?

$$\overset{?}{\frac{4}{12} = \frac{13}{39}} \qquad \overset{?}{\frac{3a^2}{7a} = \frac{6a}{14a^2}}$$

$$\frac{1}{3} = \frac{1}{3} \qquad \frac{3a}{7} \neq \frac{3}{7a}$$

yes   no

In a proportion, the product of the means equals the product of the extremes. In the proportion $a:b=c:d$, $a \cdot d = b \cdot c$.

Example 1

Find the fourth term in the proportion $\frac{2}{7} = \frac{12}{x}$.

Analysis

The product of the means equals the product of the extremes.

Work

Multiply by 7x: $\quad 7x\left(\dfrac{2}{7} = \dfrac{12}{x}\right)$

$$2x = (7)(12)$$
$$2x = 84$$

Divide by 2: $\quad\quad\quad x = 42$

Check

$$\frac{2}{7} = \frac{12}{x}$$

Answer

42 $\qquad\qquad \dfrac{2}{7} = \dfrac{12}{42}$ ✓

Exercises

1a. Find the third term in the proportion $3:11 = x:99$.

1b. Find the second term in the proportion $35:x = 5:9$.

Example 2

If seven pencils cost $1.68, how much will 18 pencils cost?

Analysis

This is a perfect candidate for a proportion solution.

Let x = the cost of 18 pencils.

Change $1.68 to 168 cents.

Work

$$\frac{\text{Pencils}_1}{\text{Cost}_1} = \frac{\text{Pencils}_2}{\text{Cost}_2}$$

$$\frac{7}{168} = \frac{18}{x}$$

Divide by 7:

$$7x = 3{,}024$$

$$x = 432$$

Check

$$\frac{7}{168} = \frac{18}{x}$$

$x = 432$:

$$\frac{7}{168} = \frac{18}{432}$$

$$\frac{1}{24} = \frac{1}{24} \checkmark$$

Answer

$4.32

2a.   If eight sodas cost $5.04, how much will 13 sodas cost?

2b.   Five pens cost $5.20. How much will 12 pens cost?

Exercises

Example 3

If 5 out of 27 people are redheads, how many redheads would you expect in a population of 1,620?

Analysis

Let x = the number of redheads in a population of 1,620.

Work

$$\frac{\text{Redheads}_1}{\text{Population}_1} = \frac{\text{Redheads}_2}{\text{Population}_2}$$

$$\frac{5}{27} = \frac{x}{1{,}620}$$

Divide by 27:

$$27x = 8{,}100$$

$$x = 300$$

Check

$$\frac{5}{27} = \frac{x}{1{,}620}$$

$x = 300$:

$$\frac{5}{27} = \frac{300}{1{,}620}$$

$$\frac{5}{27} = \frac{5}{27} \checkmark$$

Answer

    300 redheads

Exercises

3a. In a survey, 3 out of 200 automobiles are being serviced on any day. Out of 5,000 automobiles, how many cars would you expect to be out of service?

3b. Haworth Publishing prints hardbound as well as softbound books. Eight out of 23 books published are softbound. If the company published 207 books last year, how many softbound books were published?

    In the proportion $a : b = b : d$, b is called the **mean proportional**.

Example 4

    Find the mean proportional between 9 and 25.

Analysis

    Let 9 = the first term in the proportion.

    Let x = the second and third terms in the proportion (the means).

    Let 25 = the fourth term in the proportion.

Work

$$\frac{9}{x} = \frac{x}{25}$$

$$x^2 = 225$$

$$x = 15$$

Check

$$\frac{9}{x} = \frac{x}{25}$$

$x = 15:$ $\quad \dfrac{9}{15} = \dfrac{15}{25}$

$$\frac{3}{5} = \frac{3}{5} \checkmark$$

Answer

15

4a. Find the mean proportional between 9 and 16.

4b. Find the mean proportional between 4 and 49.

Exercises

9.3 Direct Variation

If we have two variables and both increase or decrease in the same ratio, we say that the two variables **vary directly**. For example, the length of a stretched spring, L, varies directly as the weight of the suspended object, W.

An experiment was performed in which various weights were hung from a spring. The weights, W, and the lengths, L, of the stretched spring were recorded, and a graph of the same data was drawn.

Weight, W (ounces)	Length L, (inches)
5.0	1.0
6.0	1.2
7.0	1.4
8.0	1.6
9.0	1.8
10.0	2.0

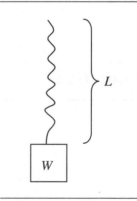

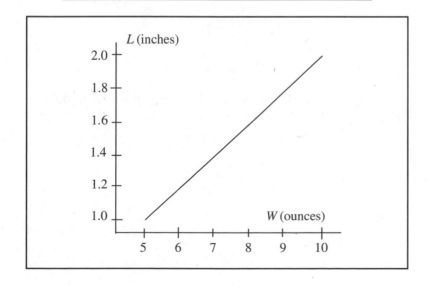

The length of the stretched spring, L, varies directly as the weight of the suspended object, W:

$$\text{Length} = \text{Constant} \times \text{Weight}$$

$$L = kW$$

The constant of the spring, k, depends upon the inherent qualities of the metal as well as the design of the spring itself.

Now let's use the results of our experiment to predict the number of inches a 12-ounce spring will stretch the spring. We can solve the problem by formula or by proportions.

Formula

First determine the constant of the spring, k, by choosing any pair of values for L and W. Let $L = 1$ and $W = 5$.

$$L = kW$$

$$1 = k \cdot 5$$

$$\frac{1}{5} = k$$

Use the values $k = 1/5$ and $W = 12$ ounces to determine the length, L.

$$L = kW$$

$$L = (1/5)12$$

$$L = 2.4$$

Answer

2.4 inches

Proportions

Choose any length and its corresponding weight from the preceding table to represent L_1 and W_1. Let $L_1 = 1$, $W_1 = 5$, $L_2 = x$, and $W_2 = 12$.

$$\frac{L_1}{W_1} = \frac{L_2}{W_2}$$

$$\frac{1}{5} = \frac{x}{12}$$

In a proportion, the product of the mean equals the product of the extremes.

$$12 \cdot 1 = 5 \cdot x$$

$$12 = 5x$$

$$2.4 = x$$

Answer

2.4 inches

Example 1

A spring has a constant of $\frac{1}{3}$. If a weight is hung from the spring and it stretches the spring 4 inches, find the weight of the object.

Analysis

Let L = length of the stretched spring.

Let k = the spring's constant.

Let W = the attached weight.

Work

$$L = kW$$

$k = \dfrac{1}{3}, L = 4:$
$$4 = \frac{1}{3}W$$

Multiply by 3:
$$3\left(4 = \frac{1}{3}W\right)$$

$$12 = W$$

Symmetric Property:
$$W = 12$$

Answer

12 ounces

The length of a stretched spring, L, varies directly as the weight of the suspended object, W.

1a. The constant of a spring is 3. If an object is hung from the spring, it stretches 4 inches. Find the weight of the object.

1b. A 6-ounce weight stretches a spring 2 inches. Find the spring's constant.

Exercises

Example 2

The length of a stretched spring, L, varies directly as the weight of the suspended object, W. If a 12-ounce weight stretches a spring 3 inches, how many inches will the same spring be stretched by a 16-ounce weight?

Analysis

The spring stretches in proportion to the attached weight. Therefore, we can use proportions to solve this problem.

Let L_1 = the distance the first spring is stretched.

Let W_1 = the first weight.

Let L_2 = the distance the second spring is stretched.

Let W_2 = the second weight.

Work

$$\frac{L_1}{W_1} = \frac{L_2}{W_2}$$

$L_1 = 3$, $W_1 = 12$,

$L_2 = x$, $W_2 = 16$: $\dfrac{3}{12} = \dfrac{L_2}{16}$

Divide by 12: $12L_2 = 48$

$L_2 = 4$

Answer

4 inches

Exercises

The length of a stretched spring, L, varies directly as the weight of the suspended object, W.

2a.   A 5-ounce weight stretches a spring 4 inches. How much will the spring be stretched by an 8-ounce weight?

2b.   A 9-ounce weight stretches a spring 3 inches. If another object stretches the spring 5 inches, how much does the object weigh?

Example 3

The real estate taxes on a house, T, vary directly as the value of the house, V. What is the tax rate, k, when \$4,320 is paid on an \$108,000 house?

Analysis

Let T = the taxes.

Let k = the tax rate.

Let V = the value of the house.

Work

The taxes vary directly as the value of the house. Use the formula $T = kV$.

$$T = kV$$

$T = \$4{,}320$, $V = 108{,}000$:	$4{,}320 = k(108{,}000)$
Divide by 108,000:	$0.04 = k$
Symmetric Property:	$k = 0.04 = 4\%$

Check

$$T = kV$$

$T = 4{,}320$, $k = 0.04$,
$V = 108{,}000$: $4{,}320 = (0.04)(108{,}000)$
                          $4{,}320 = 4{,}320$ ✓

Answer

4%

3a.   The sales tax varies directly as the price of the purchase. What is the tax rate when the tax on a $250 purchase is $22.50?

3b.   The amount of alcohol in a mixture varies directly as the weight of the solution increases. If there are 6.4 ounces of alcohol in a 40-ounce solution, what is the percentage of alcohol in the solution?

The following table records the results of an experiment in which the volume of a closed tank filled with gas was compressed. The object of the experiment was to discover the relationship between the volume of an enclosed gas and its pressure.

Exercises

9.4 Inverse Variation

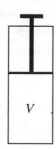

Volume, V (cu ft)	Pressure, P (lb/sq in.)
100	20
80	25
60	$33\frac{1}{3}$
40	50

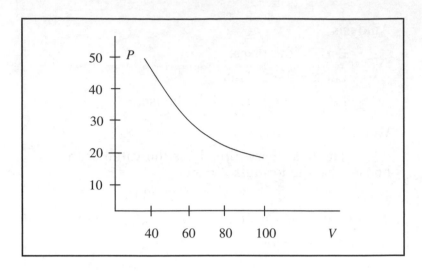

When we graph the results, they show that when the volume, V, increases, the pressure, P, decreases and vice versa. The pressure **varies inversely** with the volume.

$$PV = k$$

Let's use the results of the experiment to predict the pressure when the volume is 25 cubic feet.

We can solve the problem by formula or by proportions.

Formula

$$PV = k$$

First, find k when $P = 20$, $V = 100$:

$$20 \cdot 100 = k$$
$$2{,}000 = k$$
$$k = 2{,}000$$

Find P when $V = 25$, $k = 2{,}000$:

$$PV = k$$
$$P \cdot 25 = 2{,}000$$
$$P = 80$$

Answer

80 lb/sq in.

Proportions

$$\frac{P_1}{P_2} = \frac{V_2}{V_1}$$

$P_1 = 20$, $P_2 = ?$, $V_1 = 100$, $V_2 = 25$:

$$\frac{20}{P_2} = \frac{25}{100}$$
$$25P_2 = 2{,}000$$
$$P_2 = 80$$

Answer

80 lb/sq in.

If we have two variables and one variable moves in a direction opposite to the second variable in opposite ratios, we have a case of *inverse variation*. In physics, Boyle's Law states: At a constant temperature, the volume of gas, V, is inversely proportional to the pressure, P, applied to the gas. The greater the pressure, the smaller the volume. The lesser the pressure, the greater the volume. In other words,

$$\frac{P_1}{P_2} = \frac{V_2}{V_1}$$

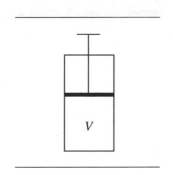

We can also say that the product of the pressure, P, and the volume, V, equals a constant, k:

$$PV = k$$

Example 1

At a constant temperature, the volume of gas, V, is inversely proportional to the pressure, P, applied to the gas. A closed tank contains 300 cubic feet of gas at an unchanging temperature. If the constant is 6,000, find the pressure in the tank.

Analysis

Let P = pressure.

Let V = the volume.

Let k = the constant.

Use the formula $PV = k$.

Work

$$PV = k$$

$V = 300$, $k = 6{,}000$: $P(300) = 6{,}000$

Divide by 300: $P = 20$

Check

$$PV = k$$

$P = 20$, $V = 300$, $k = 6{,}000$: $(20)(300) = 6{,}000$

$6{,}000 = 6{,}000$ ✓

Answer

20 lb/square in.

At a constant temperature, the volume of gas, V, is inversely proportional to the pressure, P, applied to the gas.

1a.   A closed tank contains 80 cubic feet of gas at an unchanging temperature. Find the pressure in the tank if the constant is 1,760.

1b.   The pressure in a tank is 55 pounds/square inch. Find the volume if the constant is 440.

Example 2

A group of friends decides to hire a houseboat and take it down the Mississippi. The captain has listed the following chart, indicating the cost per day per person for renting his boat. He allows a limit of 9 passengers on his boat. What is the cost *per passenger* for nine people to rent the boat?

Number of Passengers, n	1	2	3	4	5	6	9
Cost Per Passenger, c	900	450	300	225	180	150	x

Analysis

The greater the number of passengers, the less the cost per passenger.

$$1 \times 900 = 900 \qquad 4 \times 225 = 900$$
$$2 \times 450 = 900 \qquad 5 \times 180 = 900$$
$$3 \times 300 = 900 \qquad 6 \times 150 = 900$$

In all cases, the constant is the same, 900.

This sort of situation is a case of inverse variation. Although we could find the constant and then find the cost per passenger for nine people to rent the boat, let's use proportions to solve this question.

Let n_1 = the number of passengers in case 1.

Let c_1 = the cost per passenger in case 1.

Let n_2 = the number of passengers in case 2.

Let c_2 = the cost per passenger in case 2.

Work

$$\frac{n_1}{c_2} = \frac{n_2}{c_1}$$

$n_1 = 3,\ c_1 = 300,\ n_2 = 9,\ c_2 = x:$ $\qquad \dfrac{3}{x} = \dfrac{9}{300}$

Divide by 9: $\qquad\qquad\qquad\qquad\quad 9x = 900$

$$x = 100$$

Answer

$100 per passenger

Exercises

2a. The Jackson family wants to hire a computer instructor for their children for the day. If they share the expense with other families, the cost per family decreases. The table lists the *cost per family* for the indicated number of families. Find the cost per family if eight families decide to share the expenses.

Number of Families, n	2	4	5	8
Cost Per Family, c	$125	$62.50	$50	x

2b. A group of friends wants to rent a car for the day. The cost of the rental is fixed. If they share the expenses, the *cost per person* goes down. The following table lists the *cost per person* for the indicated number of people. If the cost per person is reduced to $30, how many people need to rent the car?

Number of Passengers, p	2	3	4	x
Cost Per Passenger, c	$105	$70	$52.50	$30

The Law of the Lever

We have a balance scale and two objects at different distances from the fulcrum. The law of the lever holds that if a scale is balanced by two objects on either side of a fulcrum, the distance of each object away from the fulcrum is inversely proportional to its weight.

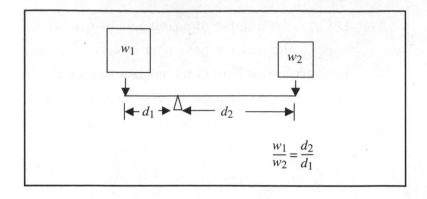

$$\frac{w_1}{w_2} = \frac{d_2}{d_1}$$

It's sometimes easier to rearrange this equation by multiplying the means and the extremes:

$$w_1 \cdot d_1 = w_2 \cdot d_2$$

Example 3

If a scale is balanced by two objects on either side of a fulcrum, the distance of each object away from the fulcrum is inversely proportional to its weight. If a 3-ounce weight placed 10 inches away from the fulcrum balances a 5-ounce weight on the other side of the fulcrum, how far from the fulcrum should the 5-ounce weight be placed?

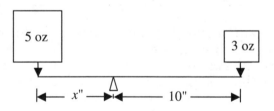

Analysis

Let w_1 = the first weight.

Let d_1 = the distance of the first weight from the fulcrum.

Let w_2 = the second weight.

Let d_2 = the distance of the second weight from the fulcrum.

We have a case of inverse variation.

Work

$$w_1 \cdot d_1 = w_2 \cdot d_2$$

$w_1 = 3, d_1 = 10, w_2 = 5, d_2 = x$: $3(10) = 5x$

$$30 = 5x$$

Divide by 5: $6 = x$

Symmetric Property: $x = 6$

Answer

6 inches

If a scale is balanced by two objects on either side of a fulcrum, the distance of each object away from the fulcrum is inversely proportional to its weight.

Exercises

3a.   A 12-ounce weight is placed 6 inches away from a fulcrum. If an 8-ounce weight is placed on the other side of the fulcrum and balances the 12-ounce weight, how far away from the fulcrum must it be placed?

3b.   A 14-ounce object is placed 8 inches away from the fulcrum. Another object is placed 16 inches away from the fulcrum on the other side. How much must this second object weigh to balance the 14-ounce weight?

Example 4

Moesha is driving a car. The time it takes for her to reach her destination is inversely proportional to her rate of speed. If she drives at 60 mph, it takes her six hours to reach her destination. How long will it take for her to reach the same destination from the same starting point if she drives at 40 mph?

Analysis

Let r_1 = the first rate of speed.

Let t_1 = the time it takes to reach the destination driving at the first rate of speed.

Let r_2 = the second rate of speed.

Let t_2 = the time it takes to reach the destination driving at the second rate of speed.

Use inverse proportions.

Work

$$\frac{r_1}{r_2} = \frac{t_2}{t_1}$$

$r_1 = 60,\ t_1 = 6,\ r_2 = 40,\ t_2 = x:$ $\dfrac{60}{40} = \dfrac{x}{6}$

$$360 = 40x$$

Divide by 40: $9 = x$

Symmetric Property: $x = 9$

Answer

9 hours

Exercises

4a.   A group of friends is planning a trip. The cost per student is inversely proportional to the number of students. If 25 students take the trip, the cost is $6 per student. Find the cost per student if 30 students take the trip.

4b.   The time it takes to complete landscaping a lawn is inversely proportional to the number of laborers. If it takes three men 12 hours to complete the job, how long would it take four men to complete the same job?

9.5 Joint Variation

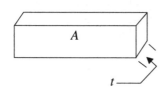

Joint variation is direct variation involving three variables.

x varies jointly as y and z: $x = kyz$

The weight of a rectangular solid, w, varies jointly as its area, A, and its thickness, t.

$$w = kAt$$

In this case, the constant, k, represents the weight per cubic inch.

Example 1

In the preceding illustration, the weight, w, varies jointly as the area, A, of the object and its thickness, t. If $k = 2$ ounces per cubic inch, $A = 23$ square inches, and $t = 0.5$ inches, find the weight of the object.

Analysis

Let w = weight.

Let k = the constant.

Let A = the area.

Let t = the thickness of the metal.

Work

$k = 2$, $A = 23$, $t = 0.5$:

$$w = kAt$$
$$w = 2(23)(0.5)$$
$$w = 23$$

Answer

23 ounces

1a. The cost of manufacturing a car battery, b, varies jointly as the cost of labor, l, and the cost of the raw materials, r. If the raw materials cost \$6, the cost of labor is \$2, and the constant, k, is 0.2, find the cost, in dollars, of manufacturing the battery.

1b. The weight of a car, wt, varies jointly as the length, l, and width, wd, of the model. If a car is 14 feet long, 6 feet wide, and the constant is 80, find the car's weight, in pounds.

Example 2

The number of bushels of corn harvested per acre, b, varies jointly as the number of inches of rain, r, and the number of pounds of seed, s, planted. If 650 bushels of corn were harvested from one acre, 2.6 inches of rain fell, and 20 pounds of seed were planted, find the constant.

Analysis

Let b = the number of bushels harvested.

Let r = the number of inches of rain.

Let s = the number of pounds of seed planted.

Let k = the constant.

Work

$$b = krs$$

$b = 650, r = 2.6, s = 20$:
$$650 = k\,(2.6)(20)$$
$$650 = 52k$$
$$k = 12.5$$

Answer

$$k = 12.5$$

Exercises

2a. The amount of garbage, g, generated monthly in an apartment house varies jointly as the number of tenants, t, and their average weight, a. If the apartment house generates 1,087.5 pounds of garbage in a month, there are 25 tenants, and the average weight per tenant is 145 pounds, find the constant, k.

2b. A company's sales, s, varied jointly as the number of salespeople, p, and the amount of advertising, a. The company's sales amounted to $270,000. If they had six salespeople and spent $50,000 on advertising, what was the constant, k?

9.6 Combined Variation

Combined variation is a combination of direct (or joint) variation and inverse variation.

y varies directly as x and inversely as z:

$$y = \frac{kx}{z}$$

w varies jointly as x and y and inversely as z:

$$w = \frac{kxy}{z}$$

Copper wire is a good conductor (carrier) of electricity. The resistance (ohms) to the electrical current in the conductor depends upon the material of the conductor (a constant), its length, and its cross-sectional area.

Example 1

The electrical resistance in a conductor, R, varies directly as the length of the conductor, L (in centimeters), and inversely as its cross-sectional area, A (in square centimeters). Determine the resistance (ohms) of a conductor if its length is 4 centimeters, its constant is 6, and its cross-sectional area is 30 square centimeters.

Analysis

Let R = the resistance.

Let k = the constant.

Let L = the length of the conductor.

Let A = the cross-sectional area.

The resistance varies directly as the length and inversely as the area.

Work

$$R = \frac{kL}{A}$$

$k = 6$, $L = 4$, $A = 30$: $\qquad R = \dfrac{6 \cdot 4}{30} = \dfrac{24}{30} = 0.8$

Answer

$R = 0.8$

The electrical resistance in a conductor, R, varies directly as the length of the conductor, L, and inversely as its cross-sectional area, A.

Exercises

1a.   Determine the resistance (in ohms) of a conductor with a cross-sectional area of 12 square centimeters, a length of 18 centimeters, and a constant of 4.

1b.   Find the resistance (in ohms) of a piece of copper wire 20 centimeters long and 2 square centimeters in cross-sectional area, with a constant of 0.3.

Example 2

The electrical resistance in a conductor, R, varies directly as the length of the conductor, L (in centimeters), and inversely as its cross-sectional area, A (in square centimeters). Find the cross-sectional area of a conductor whose length is 10 centimeters, resistance is 20 ohms, and constant is 6.

Analysis

Let R = the resistance.

Let k = the constant.

Let L = the length of the conductor.

Let A = the cross-sectional area of the conductor.

The resistance varies directly as the length of the conductor and inversely as the area.

Work

$$R = \frac{kL}{A}$$

$R = 20$, $k = 6$, $L = 10$:

$$20 = \frac{6 \cdot 10}{A}$$

Multiply by A:

$$A\left(20 = \frac{60}{A}\right)$$

$$20A = 60$$

Divide by 20:

$$A = 3$$

Answer

$A = 3$ square centimeters

Exercises

The electrical resistance in a conductor, R, varies directly as the length of the conductor, L, and inversely as its cross-sectional area, A.

2a.   Find the cross-sectional area of an electrical wire whose resistance is 35 ohms, constant is 0.7, and length is 30 centimeters.

2b.   Find the length of a copper wire whose cross-sectional area is 0.6 square centimeters, constant is 0.7, and resistance is 28 ohms.

Test

English Units	Metric Units
Distance	
1 foot = 12 inches	1 meter = 100 centimeters
1 yard = 3 feet	1 meter = 1,000 millimeters
1 yard = 36 inches	1 kilometer = 1,000 meters
Mass	
1 pound = 16 ounces	1 kilogram = 1,000 grams
1 ton = 2,000 pounds	
Volume	
1 quart = 2 pints	1 liter = 1,000 milliliters
1 gallon = 4 quarts	

1–3. Reduce to simplest terms.

1. $72a^5 : 16a^3$

2. 5 yards: 4 feet

3. 6 gallons: 3 quarts

4. Out of a sample of 300 television viewers, 40 watched the 11 o'clock news. In simplest terms, what was the ratio of the number of people who didn't watch the 11 o'clock news to the entire sample?

5. Vinegar and oil are to be mixed in a salad dressing in the ratio of 2:5, with oil in the larger proportion. If there are 63 ounces in the dressing, how many ounces of oil are there?

6. The ratio of freshmen to sophomores is in the ratio of 5:4. If 45 new freshmen enter, there will be twice as many freshmen as original sophomores. How many freshmen are there right now?

7. Find the first term in the proportion $x:19 = 55:95$.

8. If 14 light bulbs cost $10.08, how much will 17 light bulbs cost?

9. Find the mean proportional between 4 and 64.

10–11. The length of a stretched spring, L (inches), varies directly as the weight of the suspended object, W (ounces).

10. The constant of a spring is two. If an object is hung from the spring, it stretches 3 inches. Find the weight of the object.

11. A 4-ounce weight stretches a spring 1.2 inches. Find the spring's constant.

12. A salesperson's commission varies directly as her sales. If she sold $670 worth of merchandise and received a commission of $53.60, what is the rate of commission?

13–14. The pressure of enclosed gas at a constant temperature varies inversely with its volume.

13. The volume of a gas in an enclosed tank at a constant temperature is 180 cubic feet. If the constant is 6,120, find the pressure.

14. If the volume of an enclosed gas is 200 cubic feet and the pressure is 31 pounds per square inch, find the constant.

15. A group of friends decide to rent a boat for the day. The following table lists the rental price *per person* for various numbers of people. Find the rental price per person if 10 people decide to share the rental.

Number of People, n	5	6	8	10
Cost Per Person, c	$60	$50	$37.50	x

16. If a scale is balanced by two objects on either side of a fulcrum, the distance of each object away from the fulcrum is inversely proportional to its weight. A 15-ounce weight is placed 4 inches away from a fulcrum. If a 20-ounce weight is placed on the other side of the fulcrum and balances the 15-ounce weight, how far away from the fulcrum must it be placed?

17. The cost of manufacturing a transistor radio, c, varies jointly as the cost of parts, p, and the cost of labor, l. It costs $8 to manufacture the radio. If the parts cost $4 and the labor costs $0.80, what is the constant?

18. The number of apples harvested in an orchard varies jointly as the average number of branches on a tree, A, and the number of trees, t. If an orchard has 200 trees, the average number of branches is 15 and the constant is 8, how many apples are harvested?

19–20. The electrical resistance in a conductor, R, varies directly as the length of the conductor, L, and inversely as its cross-sectional area, A.

19. Determine the resistance (in ohms) of a conductor with a cross-sectional area of 16 square centimeters, a length of 25 centimeters, and a constant of 2.

20. Find the cross-sectional area of an electrical wire whose resistance is 24 ohms, constant is 4, and length is 30 centimeters.

10
Quadratic Equations

Example 1

Translate the following information into an equation. I'm thinking of a number. If we square the number, reduce it by five times the number and then add 6, the result equals zero.

Analysis

Let x = the number.

Let x^2 = the number squared.

Let $5x$ = five times the number.

Work

The square of a number reduced by five times the number added to 6 equals zero:

$$x^2 - 5x + 6 = 0$$

Answer

$$x^2 - 5x + 6 = 0$$

Exercises

Let x = the number and translate the given information into equations.

1a. Eight more than seven times a number is 22.

1b. Four less than six times a number is 67.

1c. Nine times a number reduced by 11 equals 56.

1d. 83 reduced by five times a number equals 23.

A **quadratic equation** in one unknown is an equation in which the highest power of the unknown is the second power. In the general quadratic equation $ax^2 + bx + c = 0$, a, b, and c are real numbers and $a \neq 0$. a and b are known as the coefficients in the equation, and c is the constant.

Quadratic equations are widely used to solve all sorts of problems, including

● Consumer problems

● Number problems

● Area problems

● Gravitation problems

● Work problems

● Pythagorean Theorem problems

● Electrical problems

The following equations are various forms of quadratic equations. Very often, a quadratic equation is all scrambled up, so first we need to rearrange it in descending order of exponents in the form of $ax^2 + bx + c = 0$. At that point, we can determine the coefficients of x^2 and x (a and b) as well as the constant (c).

Quadratic Equation	a	b	c
$ax^2 + bx + c = 0$	a	b	c
$3x^2 + 4x + 9 = 0$	3	4	9
$-\frac{5}{8}x^2 + 7x + 4 = 0$	$-\frac{5}{8}$	7	4
$+\frac{2}{3}t^2 + 2t \quad = 0$	$+\frac{2}{3}$	2	0

Example 2

In the equation $8 - x = 7x^2$, rearrange the terms in descending order of exponents and determine the coefficients of x^2 and x (a and b) as well as the constant (c).

Analysis

The generic quadratic equation is in the form $ax^2 + bx + c = 0$. Let's arrange the equation in this form.

Work

$$8 - x = 7x^2$$

Rearrange the equation
in descending order of
exponents: $\qquad 0 = 7x^2 + x - 8$

Symmetric Property: $\qquad 7x^2 + x - 8 = 0$

Generic Format: $\qquad ax^2 + bx + c = 0$

Answer

$\qquad a = 7, b = 1, c = -8$

Exercises

In the following equations, arrange the equations in the form $ax^2 + bx + c = 0$ and then determine a, b, and c.

2a. $+7x^2 + 3x - 8 = 0$ 2b. $-9x^2 - x - 9 = 0$

2c. $4x^2 = 0$ 2d. $5x^2 + 6 = 0$

2e. $5 = 8x - 2x^2$ 2f. $13 - 2x = 4x^2$

2g. $3x + 6x^2 - 1 = 9$ 2h. $5 - 3x^2 = x^2 + 4x$

2i. $4x(x - 3) = 9$ 2j. $-3(x^2 + 7x) - 6 = 0$

10.2 Finding the Roots of a Quadratic Equation

Let's say that we have the quadratic equation $x^2 - 5x + 6 = 0$ and we want to find the variable that will make this equation true. We can try substituting various numbers for x, but I happen to know that one answer is 2, so let's see what happens when we substitute 2 for x in the equation.

$$x^2 - 5x + 6 = 0$$
$$(2)^2 - 5(2) + 6 = 0$$
$$4 - 10 + 6 = 0$$
$$10 - 10 = 0 \checkmark$$

I happen to know that another answer is 3, so let's try substituting that now:

$$x^2 - 5x + 6 = 0$$
$$(3)^2 - 5(3) + 6 = 0$$
$$9 - 15 + 6 = 0$$
$$15 - 15 = 0 \checkmark$$

A quadratic equation has two answers or **roots**. In this illustration, 3 and 2 are the answers or roots of the quadratic equation. If we substitute a root into an equation, we obtain a **balanced** or a **true equation**.

There are a number of methods for finding the roots of a quadratic equation. However, we'll restrict ourselves to the three most widely used techniques:

1. Drawing a graph

2. Factoring

3. Using a formula

We'll start solving quadratic equations by drawing a graph because this method will give us the most insight into quadratic equations. As the saying goes, "A picture is worth a thousand words."

Finding the Roots by Drawing a Graph

Example 1

Find the roots of the quadratic equation $12 = (2x + 2)x$.

Analysis

Simplify the equation and graph it.

Work

$$12 = x(2x + 2)$$

Multiply factors on right-hand side:

$$12 = 2x^2 + 2x$$

Subtract 12:

$$0 = 2x^2 + 2x - 12$$

Divide by 2:

$$0 = x^2 + 1x - 6$$

Symmetric Property:

$$x^2 + 1x - 6 = 0$$

We can't draw a graph of this quadratic equation because we can substitute only two numbers for x that can satisfy the equation. We're going to introduce one change into the equation and substitute y for 0: $y = x^2 + 1x - 6$. This slight change will permit us to determine numerous values for y corresponding to different values for x. We'll thus have a larger set of values for x and y, which will permit us to graph the equation on the coordinate plane.

After we have a graph, we'll see where the graph crosses the x-axis. At these points, $y = 0$, so we're back to our original equation, $0 = x^2 + 1x - 6$ and then we'll see what x is at these points.

Let's sketch the graph for the domain $-3 \leq x \leq +3$.

$$y = x^2 + 1x - 6$$

x	x^2	$+$	x	$-$	6	y	(x, y)
-3	$(-3)^2$	$+$	(-3)	$-$	6	0	$(-3, 0)$
-2	$(-2)^2$	$+$	(-2)	$-$	6	-4	$(-2, -4)$
-1	$(-1)^2$	$+$	(-1)	$-$	6	-6	$(-1, -6)$
0	$(0)^2$	$+$	(0)	$-$	6	-6	$(0, -6)$
1	$(1)^2$	$+$	(1)	$-$	6	-4	$(1, -4)$
2	$(2)^2$	$+$	(2)	$-$	6	0	$(2, 0)$
3	$(3)^2$	$+$	(3)	$-$	6	6	$(3, 6)$

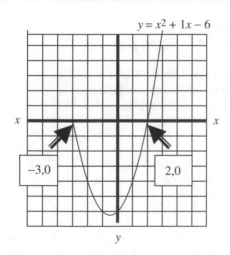

$y = x^2 + 1x - 6$

$-3,0$

$2,0$

We're looking for the point(s) along the graph where $y = 0$. If we look carefully at this graph, we can see that the curve $y = x^2 + x - 6$—a parabola—crosses the x-axis at the points $(-3, 0)$ and at $(2, 0)$. These are the two points that satisfy the equation $0 = x^2 + x - 6$. Note that $y = 0$ at both these points.

Check

Let's go back to the original equation and check both answers.

$x = -3$	$x = 2$
$12 = (2x + 2)x$	$12 = (2x + 2)x$
$12 = \{(2)(-3) + 2\}(-3)$	$12 = \{2(2) + 2\}(2)$
$12 = \{-6 + 2\}(-3)$	$12 = \{4 + 2\}(2)$
$12 = \{-4\}(-3)$	$12 = \{6\}(2)$
$12 = 12$ ✓	$12 = 12$ ✓

Answer

$x = -3, 2$

Exercises

Find the roots of these quadratic equations by drawing a graph.

1a. $x^2 - 4x + 3 = 0$ 1b. $x^2 = 8 - 2x$

1c. $2x^2 - 2x - 12 = 0$ 1d. $-3x^2 = -12 - 9x$

Finding the Roots by Factoring

In the majority of cases, the most convenient method for finding the roots of a quadratic equation is by factoring. Let's say that we're given the statement "the product of a and b is zero." For this statement to be true, either a or b or both numbers must be zero.

Example 2

Let's take the quadratic equation that we looked at earlier, namely $12 = (2x + 2)x$, factor it, set it equal to 0 and try to solve it.

Analysis

Set the equation equal to 0. Then factor and set each of the factors equal to zero.

Work

$$12 = (2x + 2)x$$

Symmetric Property: $(2x + 2)x = 12$

Multiply the factors on the left: $2x^2 + 2x = 12$

Subtract 12: $2x^2 + 2x - 12 = 0$

Divide by 2: $x^2 + 1x - 6 = 0$

Factor: $(x + 3)(x - 2) = 0$

Set each factor equal
to zero:

$$x + 3 = 0 \qquad x - 2 = 0$$

$$x = -3 \qquad x = +2$$

These answers agree with the preceeding solution where we found the roots by graphing the quadratic equation.

Answer

$$x = -3, 2$$

Determine the roots in these quadratic equations by factoring.

2a. $x^2 + 5x + 6 = 0$ 2b. $y^2 - y - 6 = 0$

2c. $5z^2 = 125$ 2d. $3r^2 = 8 + 10r$

In case we can't easily factor a quadratic equation, we can use a formula to find the roots of the equation. Mathematicians have developed the quadratic formula that allows us to simply "plug in" numbers for a, b, and c and then grind out the answers.

Finding the Roots by Formula

Example 3

Let's go back to our original equation, $12 = (2x + 2)x$, and find the roots using the quadratic formula. According to the formula,

$$x_1 = \frac{-b + \sqrt{b^2 - 4ac}}{2a} \quad \text{and} \quad x_2 = \frac{-b - \sqrt{b^2 - 4ac}}{2a}$$

Analysis

Arrange the equation in descending order of exponents and set it equal to zero. Find the coefficients of x^2 and x, a and b, and the constant, c. Then substitute these values directly into the quadratic formula above.

Work

$$(2x + 2)x = 12$$

Multiply the factors
on the left:

$$2x^2 + 2x = 12$$

Subtract 12 from
both sides:

$$2x^2 + 2x - 12 = 0$$

We could use the coefficients of x^2 and x and the constant and substitute these values into the quadratic formula. However, to simplify our work, it is often easier to first simplify the equation and then to substitute the newer values.

Divide both sides by 2: $\quad x^2 + 1x - 6 = 0$
$$a = 1,\ b = 1,\ c = -6$$

Substitute these values into the formula:

$$x = \frac{-b \pm \sqrt{b^2 - 4ac}}{2a} = \frac{-(1) \pm \sqrt{(1)^2 - 4(1)(-6)}}{2(1)}$$

$$= \frac{-1 \pm \sqrt{1 + 24}}{2} = \frac{-1 \pm \sqrt{25}}{2} = \frac{-1 \pm 5}{2}$$

There are two roots to the quadratic equation:

$$x_1 = \frac{-1 + 5}{2} = \frac{+4}{2} = 2$$

$$x_2 = \frac{-1 - 5}{2} = \frac{-6}{2} = -3$$

Once again, these results confirm our earlier answers.

Answer

$$x = -3,\ 2$$

Exercises

Use the quadratic formula to find the roots of the following equations.

3a. $\quad x^2 - 3x - 10 = 0$ 3b. $\quad x^2 + 3x = 18$

3c. $\quad 12x^2 = 3 - 5x$ 3d. $\quad 5x^2 - 12x - 9 = 0$

Word Problems

    We can use any of the methods sketched out in Section 10.2 to solve quadratic equations. Sometimes one method might be somewhat more convenient than another, but there is no one prescribed technique. However, no matter which procedure we use, we should all arrive at the same answer.

    In manufacturing problems, we usually use the formula:

Number of units produced *times* Cost per unit
   = Total cost

10.3 Consumer Problems

Example 1

    The Gidget Manufacturing Company can manufacture a certain number of gidgets for $900. If they manufacture 150 more gidgets, their cost drops $1 per gidget, while their total manufacturing cost remains the same. What was the original price of each gidget and how many were originally manufactured? (Hint: Factor to find the solution.)

Analysis

Let n = original number of gidgets manufactured.

Let m = original manufacturing cost per gidget.

Let $n + 150$ = the new number of gidgets manufactured.

Let $m - 1$ = the new manufacturing cost per gidget.

Use the general formula: Number of units × Manufacturing cost per unit = Total manufacturing cost

Work

The company can manufacture a certain number of gidgets for $900:

$$(1) \quad n \cdot m = 900$$

If they manufacture 150 more gidgets, their cost drops $1 per gidget while their total manufacturing cost remains the same:

$$(2) \quad (n + 150)(m - 1) = 900$$

Let's solve for n in our original equation:

$$(1) \quad nm = 900$$

Divide by m:

$$(1) \quad n = \frac{900}{m}$$

Substitute this value of n into the second equation:

$$(2) \quad (n + 150)(m - 1) = 900$$

$n = 900/m$:

$$(2) \quad \left(\frac{900}{m} + 150\right)(m - 1) = 900$$

Multiply on the left side:

$$(2) \quad 900 - \frac{900}{m} + 150m - 150 = 900$$

Multiply by m:

$$(2) \quad m\left(900 - \frac{900}{m} + 150m - 150\right) = 900$$

$$(2) \quad 900m - 900 + 150m^2 - 150m = 900m$$

$$(2) \qquad\quad 750m - 900 + 150m^2 = 900m$$

$$(2) \quad 750m - 900 + 150m^2 - 900m = 0$$

$$(2) \qquad\quad -150m - 900 + 150m^2 = 0$$

Rearrange terms in
descending order
of exponents: $\quad (2) \quad 150m^2 - 150m - 900 = 0$

Divide by 150: $\quad (2) \qquad\qquad m^2 - m - 6 = 0$

Factor: $\qquad\qquad (2) \qquad\qquad (m - 3)(m + 2) = 0$

Set each factor equal
to zero: $\qquad\qquad\qquad m - 3 = 0 \qquad m + 2 = 0$

$$m = 3 \qquad\quad m = -2$$

Reject. The cost cannot be negative.

To find n,
substitute $m = 3$ into the
original equation: $\qquad (1) \quad n \cdot m = 900$

$$(1) \quad n \cdot 3 = 900$$

$$(1) \qquad n = 300$$

Check

$$(1) \qquad\qquad\qquad n \cdot m = 900$$

$$(1) \qquad\qquad\qquad 300 \cdot 3 = 900 \checkmark$$

$$(2) \quad (n + 150)(m - 1) = 900$$

$$(2) \quad (300 + 150)(3 - 1) = 900$$

$$(2) \qquad\qquad\qquad 450 \cdot 2 = 900$$

$$(2) \qquad\qquad\qquad 900 = 900 \checkmark$$

Answer

The Gidget Manufacturing Company
originally manufactured 300 units at $3 per unit.

Exercises

1a. Manny got paid $1,120 a week at his job. If he received a $4 per hour increase in salary, he would be able to work 5 hours a week less and receive the same total wages. What was his original hourly wage?

1b.   A pump can fill a storage tank with a capacity of 3,000 gallons of oil. A new pump can fill the tank in one hour less time if it increases the pumping rate by 100 gallons per hour. How many hours would it take for the original pump to fill the tank?

1c.   The Skoda Car Company can manufacture 144 cars in a certain number of days. If the production rate is increased by four units per day, the time it takes to complete 144 units is decreased by six days. How many cars per day did Skoda manufacture at the original rate?

1d.   A secretary is shared by a certain number of lawyers. Collectively, they pay him $480 per week. If four more lawyers join in the pool and the secretary still receives the same total weekly salary, each lawyer can reduce her payments by $20 per week. How much did each member of the original group chip in to pay the secretary?

Example 1

Two times the square of an integer decreased by three times the integer is 20. Find the integer by factoring.

Analysis

Let x = the integer.

Let x^2 = the square of the integer.

Work

Two times the square of an integer decreased by three times the integer is 20:

$$2x^2 - 3x = 20$$

Subtract 20: $2x^2 - 3x - 20 = 0$

Factor: $(2x + 5)(x - 4) = 0$

Set each factor equal to zero:

$$2x + 5 = 0 \qquad x - 4 = 0$$
$$2x = -5 \qquad\qquad x = 4$$
$$x = -2\frac{1}{2}$$

Reject because $-2\frac{1}{2}$ is not an integer.

Check

Go back to the original equation.

$$2x^2 - 3x = 20$$

$$2(4)^2 - 3(4) = 20$$
$$2(16) - 12 = 20$$
$$32 - 12 = 20$$
$$20 = 20 \checkmark$$

Answer

4

Exercises

1a. If six times the square of an integer is decreased by eight times the integer, the result is 8. Find the integer.

1b. If two times the square of an integer is decreased by three times the integer, the result is 77. Find the positive integer.

1c. The square of a positive integer is 16 more than six times the integer. Find the positive integer.

1d. Four times the square of an integer decreased by seven times an integer equals 15. Find the integer.

Consecutive integers are integers that follow each other without a break in order. For example, if 45 is the first integer, 46 is the next consecutive integer. Similarly, if we let x represent the first integer, then $x + 1$ represents the next consecutive integer.

Consecutive even integers are even integers that follow each other without a break in order. For example, if 22 is the first even integer, 24 is the next consecutive even integer. If x represents the first even integer, then $x + 2$ represents the next consecutive even integer.

Consecutive odd integers are odd integers that follow each other without a break in order. For example, if 77 is the first odd integer, 79 is the next consecutive odd integer. If x represents the first odd integer, then $x + 2$ represents the next consecutive odd integer.

Example 2

The product of two consecutive positive odd integers equals 91 more than eight times the first integer. Find both integers.

Analysis

Let x = the first odd integer.

Let $x + 2$ = the next consecutive odd integer.

Work

The product of two consecutive positive odd integers equals 91 more than 8 times the first integer:

$$x(x + 2) = 8x + 91$$

Multiply the factors: $x^2 + 2x = 8x + 91$

Subtract 8x and 91: $x^2 + 2x - 8x - 91 = 0$

Simplify terms: $x^2 - 6x - 91 = 0$

Factor: $(x - 13)(x + 7) = 0$

Set the factors equal to zero:

$$x - 13 = 0 \qquad x + 7 = 0$$

$$x = 13 \qquad x = -7$$

$x + 2 = 15$ Reject because –7 not a positive integer.

Check

$$x(x + 2) = 8x + 91$$
$$(13)(15) = 8(13) + 91$$
$$195 = 104 + 91$$
$$195 = 195 \checkmark$$

Answer

The two consecutive positive odd integers are 13 and 15.

2a. If the product of two positive consecutive integers is 132, find them.

2b. The product of two positive consecutive odd integers is 22 more than seven times the second odd integer. Find both integers.

2c. The product of two consecutive positive even integers is 10 less than nine times the second integer. Find both integers.

2d. Find two consecutive positive integers such that the sum of their squares is 113.

Exercises

10.5 Area Problems

Example 1

The base of a triangle is 4 inches more than the height. If the area is 16 square inches, find the base and the height.

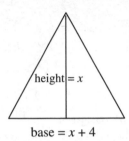

height = x

base = $x + 4$

Analysis

Let x = the height.

Let $x + 4$ = the base.

Work

$$A = \frac{1}{2} bh$$

Area = 16: $16 = \frac{1}{2}(x + 4)(x)$

Multiply by 2: $32 = (x + 4)(x)$

$$32 = x^2 + 4x$$

$$0 = x^2 + 4x - 32$$

Symmetric Property: $x^2 + 4x - 32 = 0$

Factor: $(x + 8)(x - 4) = 0$

Set each factor equal to zero:

$$x + 8 = 0 \qquad x - 4 = 0$$
$$x = -8 \qquad x = 4$$

Reject because the base cannot be −8. $x + 4 = 8$

Check

$$A = \frac{1}{2}bh$$

$$16 = \frac{1}{2} \times 8 \times 4$$

$$16 = \frac{1}{2} \times 32$$

$$16 = 16$$

Answer

The height, x, is 4; the base, $x + 4$, is 8.

1a.   The area of a triangle is 33 square inches. If the height is 1 inch less than twice the base, find the base and height of the triangle.

1b.   The base of a triangle is 2 inches more than three times the height. If the area of the triangle is 28 square inches, find the base of the triangle.

Example 2

The base of a rectangle is 3 less than twice the height. If the area is 135 square inches, find the dimensions of the rectangle.

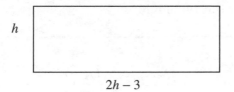

$2h - 3$

Analysis

Let h = the height.

Let $2h - 3$ = the base.

Work

$$A = bh$$
$$135 = (2h - 3)h$$

Multiply factors: $135 = 2h^2 - 3h$

Subtract 135: $0 = 2h^2 - 3h - 135$

Symmetric
Property: $2h^2 - 3h - 135 = 0$

Factor: $(h - 9)(2h + 15) = 0$

Set each factor equal to zero:

$$h - 9 = 0 \qquad 2h + 15 = 0$$
$$h = 9 \qquad 2h = -15$$
$$h = -7\frac{1}{2}$$

Reject because the height cannot be negative.

Check

$$A = bh$$
$$135 = (2h - 3) \cdot h$$
$$135 = \{2(9) - 3\} \cdot 9$$
$$135 = 15 \cdot 9$$
$$135 = 135 \checkmark$$

Answer

The height is 9; the base is 15.

Exercises

2a. The width of a rectangle is 3 inches less than its length. If the area of the rectangle is 70 square inches, find the length of the rectangle.

2b. If the width of a rectangle is 16 inches less than twice the length and the area is 96 square inches, find the width of the rectangle.

Example 3

The length of a rectangle is 2 inches more than its width. If the length is increased by 4 inches and the width is doubled, a new rectangle is formed whose area is 75 square units more than the old area. Find the dimensions of the original rectangle.

Analysis

Original Rectangle

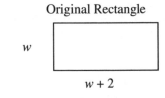

w

$w + 2$

Let w = the width.

Let $w + 2$ = the length.

Area = $w(w + 2)$.

New Rectangle

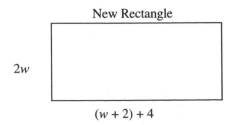

$2w$

$(w + 2) + 4$

Let $2w$ = the width.

Let $(w + 2) + 4 = w + 6$ = the length.

Area = $2w(w + 6)$.

Work

The new area is 75 square units larger than the old area:

$$(2w)(w + 6) = w(w + 2) + 75$$

$$2w^2 + 12w = w^2 + 2w + 75$$

Rearrange the terms: $w^2 + 10w - 75 = 0$

Factor: $(w + 15)(w - 5) = 0$

$$w + 15 = 0 \qquad w - 5 = 0$$

$$w = -15 \qquad w = 5$$

Reject because the width $w + 2 = 7$
cannot be negative.

Check

$$(2w)(w + 6) = w(w + 2) + 75$$

$$(2 \cdot 5)(5 + 6) = 5(5 + 2) + 75$$

$$(10)(11) = 5(7) + 75$$

$$110 = 35 + 75$$

$$110 = 110 \checkmark$$

Answer

The original rectangle is 5 inches wide and 7 inches long.

3a.   The length of a rectangle is 2 inches more than the width. If the length is increased by 4 inches and the width is doubled, a new rectangle is formed whose area is 75 square inches greater than the old area. Find the dimensions of the original rectangle.

3b.   The width of a rectangle is 3 inches less than its length. If the length is tripled and the width is decreased by 2 inches, a new rectangle is formed whose area is 144 square inches more than the area of the original rectangle. Find the length and width of the original rectangle.

Exercises

If one person can complete a job in three hours, she completes one third of the job in one hour. Conversely, if she completes one third of the job in one hour, she can complete the entire job in three

10.6 Work Problems

hours. Let's just generalize this information. We'll substitute x for 3, so that now we get the more generic statement:

If one person can complete a job in x hours, she completes $1/x$ of the job in one hour. Conversely, if she completes $1/x$ of the job in one hour, she can complete the entire job in x hours.

Let's now add a second person who can complete $1/y$ part of the job in one hour. If both work together, they complete $(1/x + 1/y)$ part of the job in one hour.

Let's say we now want to determine the number of hours, z, it will take for the two persons, working together, to complete the job.

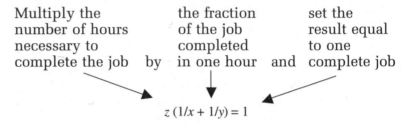

Multiply the number of hours necessary to complete the job	by	the fraction of the job completed in one hour	and	set the result equal to one complete job

$$z(1/x + 1/y) = 1$$

Example 1

Working together, Mr. Johnson and his son can complete painting a room in two hours. If it takes the son three hours longer than the father to paint the room alone, how long should it take for Mr. Johnson to paint the room alone?

Analysis

Let x = the time it takes Mr. Johnson to paint the room alone.

Let $x + 3$ = the time it takes his son to do the job alone.

If they work together, they complete $1/x + 1/(x + 3)$ part of the job in one hour. If we then multiply this part of the job by two hours, the number of hours necessary to complete the job together, we get a completed job, 1.

Work

Working together, hours necessary to complete job	Together for one hour		One complete job
2	$\left(\dfrac{1}{x} + \dfrac{1}{x+3}\right)$	$=$	1
	$\dfrac{2}{x} + \dfrac{2}{x+3}$	$=$	1

Multiply by the least common denominator:

$$(x)(x+3)\left(\frac{2}{x} + \frac{2}{x+3} = 1\right)$$

$$2(x+3) + 2(x) = x(x+3)$$

$$2x + 6 + 2x = x^2 + 3x$$

$$0 = x^2 - x - 6$$

Symmetric Property:

$$x^2 - x - 6 = 0$$

Factor:

$$(x - 3)(x + 2) = 0$$

Set each factor equal to zero:

$$x - 3 = 0 \qquad x + 2 = 0$$

$$x = 3 \qquad\qquad x = -2 \text{ Reject.}$$

$$x + 3 = 6$$

Check

$$2\left(\frac{1}{x} + \frac{1}{x+3}\right) = 1$$

$x = 3,\ x + 3 = 6$:

$$2\left(\frac{1}{3} + \frac{1}{6}\right) = 1$$

$$\left(\frac{2}{3} + \frac{2}{6}\right) = 1$$

$$\frac{6}{6} = 1$$

$$1 = 1 \checkmark$$

Exercises

1a. John can assemble a chair in 15 minutes more time than it takes Harold to assemble the same chair. If they work together, they can complete the job in 18 minutes. How long does it take each man to assemble the chair alone?

1b. Hilda takes six hours longer to complete mowing a lawn alone than it takes Vanessa to do the same job alone. If they work together, they can complete the job in four hours. How long does it take each woman to finish the job alone?

1c. It takes Leshawn three times as long to assemble a computer alone as it takes for Carmen to assemble the same computer. If the women work together, they can complete the job in six hours. How long does is take each woman working alone to complete the job?

10.7 Pythagorean Theorem Problems

    In a right triangle, the sum of the squares of the two legs equals the hypotenuse squared:

$$a^2 + b^2 = c^2$$

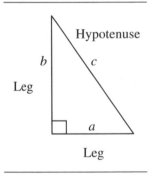

Example 1

    A ladder leans against a wall. The ladder's distance away from the base of the wall is represented by x feet, and the ladder contacts the wall at a height of $x + 14$ feet above the ground. If the ladder's length is represented by $2x + 2$, find x.

Analysis

We have a right triangle and are given the three sides in terms of x, so let's use the Pythagorean Theorem.

Let x = the ladder's distance away from the foot of the wall.

Let $x + 14$ = the height of the ladder's contact point with the wall above the ground.

Let $2x + 2$ = the length of the ladder.

Work

$$a^2 + b^2 = c^2$$
$$(x + 14)^2 + (x)^2 = (2x + 2)^2$$
$$(x + 14)(x + 14) + x^2 = (2x + 2)(2x + 2)$$
$$x^2 + 28x + 196 + x^2 = 4x^2 + 8x + 4$$
$$2x^2 + 28x + 196 = 4x^2 + 8x + 4$$
$$0 = 2x^2 - 20x - 192$$
$$0 = x^2 - 10x - 96$$

Symmetric Property:

$$x^2 - 10x - 96 = 0$$
$$(x + 6)(x - 16) = 0$$

$$x + 6 = 0 \qquad x - 16 = 0$$
$$x = -6 \qquad x = 16$$
$$\text{Reject.} \qquad x + 14 = 30$$
$$2x + 2 = 34$$

Check

$$a^2 + b^2 = c^2$$
$$30^2 + 16^2 = 34^2$$
$$900 + 256 = 1,156$$
$$1,156 = 1,156 \checkmark$$

Answer

16

Find all the missing sides in the following right triangles.

1a. Find hypotenuse c in right triangle ABC if side b is 7 more than side a and hypotenuse c is 8 more than side a.

Exercises

1b.   A city lot is in the shape of a right triangle. If the longer leg of the lot is 1 foot less than twice the shorter leg and the hypotenuse is 9 feet more than the shorter leg, find all three sides of the lot.

1c.   We want to construct a sign in the shape of a right triangle. If the longer leg of the sign is 2 inches more than the shorter leg and the hypotenuse is 2 inches less than twice the shorter leg, find all three sides of the sign.

1d.   Find the diagonal of a rectangle if the base is 14 less than the height and the diagonal is 22 less than twice the height. Hint: Use the quadratic formula.

10.8 Electrical Problems

Raw electric current as it moves from the generator through copper wiring to our homes and to industry must be manipulated before it can perform any useful tasks. Engineers have developed devices such as transistors, semiconductors, transformers, inductors, and insulators, which can manipulate the electric current to run numerous electronic devices.

The resistor is one of these useful devices engineers have developed. As its name implies, this device introduces a specific amount of resistance into the circuit.

Example 1

If there are two resistors, r_1 and r_2, in a parallel circuit, the joint resistance, R, is given by the formula

$$\frac{1}{R} = \frac{1}{r_1} + \frac{1}{r_2}$$

If $R = 4$ ohms and r_2 is 15 ohms greater than r_1, find r_1 and r_2.

Analysis

Let $r_1 = x$.

Let $r_2 = x + 15$.

Work

Substitute:
$$\frac{1}{4} = \frac{1}{x} + \frac{1}{x+15}$$

Multiply by the least common denominator, $4(x)(x + 15)$:

$$4(x)(x + 15) \quad \left(\frac{1}{4} = \frac{1}{x} + \frac{1}{x+15}\right)$$

$$x(x + 15) = 4(x + 15) + 4(x)$$

$$x^2 + 15x = 4x + 60 + 4x$$

$$x^2 + 15x = 8x + 60$$

Rearrange terms: $\quad x^2 + 7x - 60 = 0$

Factor: $\quad (x + 12)(x - 5) = 0$

Set each factor equal to zero:

$$x + 12 = 0 \qquad\qquad x - 5 = 0$$

$$x = -12 \qquad\qquad x = 5$$

$$\text{Reject.} \qquad x + 15 = 20$$

Check

$$\frac{1}{R} = \frac{1}{r_1} + \frac{1}{r_2}$$

$$\frac{1}{4} = \frac{1}{5} + \frac{1}{20}$$

$$\frac{1}{4} = \frac{25}{100}$$

$$\frac{1}{4} = \frac{1}{4} \checkmark$$

Answer

$$r_1 = 5,\ r_2 = 20$$

In the following problems, r_1 and r_2 are the two resistors in a parallel circuit, and R is the joint resistance. Use the formula

$$\frac{1}{R} = \frac{1}{r_1} + \frac{1}{r_2}$$

1a. In a parallel circuit, the joint resistance is 8 ohms. If one resistor is 12 ohms less than the second resistor, find both resistors.

1b. In a parallel circuit, one resistor is 8 ohms more than a second resistor. If the joint resistance is 3 ohms, find both resistors.

1c. If one resistor in a parallel circuit is 6 ohms and a second resistor is 1 ohm more than the joint resistance, find the number of ohms in the joint resistance.

1d. One resistor in a parallel circuit is 12 ohms. The second resistor is 2 ohms more than the joint resistance. Find the number of ohms in the second resistor.

Test

1. Draw the graph of the quadratic equation $y = x^2 - 4x + 3$ in the interval $-4 \leq x \leq 3$. Then find the roots of the equation $x^2 - 4x + 3 = 0$.

2. Find the roots of the equation $2x^2 - 3 = 5x$. Hint: Factor.

3. Use the quadratic formula to find the roots of the equation $-3x + 13 = 2x^2$. Round to the nearest tenth.

4. The High Five Beanbag Company can manufacture a certain number of beanbags per day for $180. If they manufacture 30 more beanbags per day, the manufacturing cost drops by $0.50 per unit while the total daily manufacturing cost of $180 remains the same. How many beanbags did the company originally manufacture per day?

5. The Wilton Company can manufacture a certain number of radios for $43,200. If they increase the number of radios manufactured by 900, the cost drops by $8 per unit, and the total manufacturing cost remains the same. How many radios did they manufacture originally?

6. Three times the square of a positive integer increased by four times the integer equals 224. What is the number?

7. The product of two consecutive positive integers is 75 more than nine times the second integer. Find both integers.

8. The product of the first and third of three positive consecutive odd integers is 80 less than 23 times the second integer. Find all three odd integers.

9. The product of the first and third of three positive consecutive even integers is 150 more than the sum of all three integers. Find all the integers.

10. Three-quarters times the square of a positive integer number is 3 less than five times the integer. Find the integer.

11. The square of the first consecutive integer plus the square of the third consecutive integer equals 486 more than the square of the second consecutive integer. Find the integers.

12. The area of a triangle is 108 square inches. If the height is 6 inches less than the base, find the base and the height of the triangle. Hint: Use the quadratic formula.

13. The length of a rectangle is 7 inches less than twice the width. If the area is 204 square inches, find the dimensions of the rectangle.

14. The height of a triangle is 7 more than the base. If the area equals 114, find its base and height.

15. The length of a rectangle is 3 more than the width. If the length is doubled and 4 is added to the width, a new rectangle is formed with an area 104 square units larger than the area of the original rectangle. Find the dimensions of the original rectangle.

16–17. The joint resistance, R, in a parallel electrical circuit is given by the formula:

$$\frac{1}{R} = \frac{1}{r_1} + \frac{1}{r_2}$$

where r_1 and r_2 are individual resistors.

16. If the joint resistance is 4 ohms and r_1 is 6 ohms more than r_2, find the number of ohms in each resistor.

17. If r_1 is 4 ohms more than the joint resistance, R, and r_2 is 3 ohms, find R and r_1.

18. Carlos can paint a room in three hours less time than Reggie. If both men work together, they can complete painting the entire room in two hours. How long would it take for Carlos to do the job alone?

19. A ladder leans against a wall and touches it at a point 12 feet above the ground. If the length of the ladder is 3 feet less than twice the distance from the foot of the ladder to the bottom of the wall, how long is the ladder? Hint: Use the Pythagorean Theorem.

20. In rectangle $ABCD$, diagonal $\overline{BD}$ is one more than twice $\overline{BC}$. $\overline{CD}$ is 7 more than $\overline{BC}$. Find $\overline{AB}$.

11

Trigonometry of the Right Triangle

The Greek theoretician Pythagoras (c. 569–500 B.C.E.) was a combination mystic, mathematician, and philosopher. He tried to discover the immutable laws of the universe and society, and he was certain that numbers expressed the relationship among natural phenomena. The world, in his view, was governed by harmony, and this harmony was best expressed by numerical relationships.

He founded his own school in Croton in southern Italy, and thanks to his students—appropriately enough called Pythagoreans—we have a record of his work. He and his students divided numbers into different types of classes—triangular, square, pentagonal, prime and composite, perfect, friendly, odd, even, and the like.

In his quest to discover the underlying laws of the universe, Pythagoras developed the theorem that the square of the hypotenuse of a right triangle equals the sum of the squares of its two legs—the Pythagorean Theorem. Much to their horror, the Pythagoreans soon discovered that their very own theorem contradicted the idea of a rational universe and a rational number system. When a right triangle is constructed with each leg equal to one, the hypotenuse is √2—a number that cannot be expressed as a ratio of two whole numbers. This discovery prompted the Pythagoreans to establish a code of secrecy. When one of their own members—Hippasus—revealed the flaw in their rational universe, he was put to death.

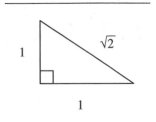

Although Pythagoras influenced later generations of Greek mathematicians such as Plato and Euclid, the Pythagoreans eventually degenerated into a cult that refused to eat beans, dabbled in vegetarianism, and believed in the transmigration of souls.

Pythagoras and his followers proposed the following theorem:

11.1 The Pythagorean Theorem

In a right triangle, the sum of the squares of the lengths of the legs equals the square of the length of the hypotenuse.

$$a^2 + b^2 = c^2$$

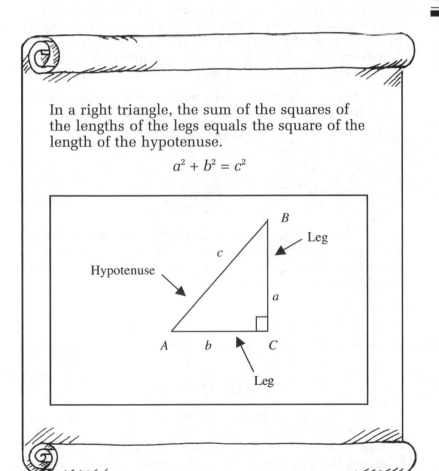

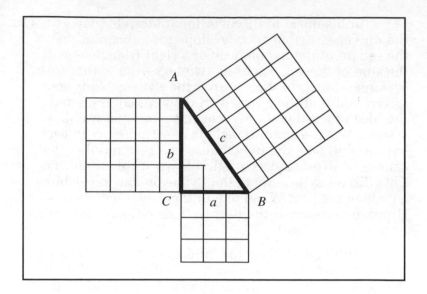

The best way to understand the Pythagorean Theorem is graphically. Right triangle ABC in the diagram has angle $C = 90°$, $\overline{BC}$ (or a) = 3, $\overline{AC}$ (or b) = 4. We have constructed three squares, one on the hypotenuse with a side equal to the length of the hypotenuse and two others on the legs with sides equal to the lengths of the legs of the triangle. The square constructed on the hypotenuse has a side of 5, and those constructed on the legs have sides of 3 and 4.

Let's got back to the Pythagorean Theorem and substitute these values.

$$a^2 + b^2 = c^2$$

$$3^2 + 4^2 = 5^2$$

$$9 + 16 = 25$$

$$25 = 25 \checkmark$$

The diagram clearly shows that the area of the square on the hypotenuse side (25) equals the sum of the areas of the squares constructed on the sides of the legs (9 + 16).

Example 1

Wei wants to cut across a rectangular lot rather than walk around it. If the lot is 120 feet long and 50 feet wide and Wei walks diagonally across the lot, how many feet is the shortcut?

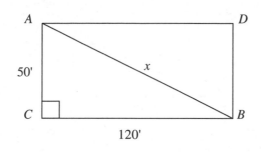

50'

120'

Analysis

We have a right triangle and we know two sides, so we have all the requirements for using the Pythagorean Theorem.

Let the diagonal = x.

Work

Pythagorean Theorem:

$$a^2 + b^2 = c^2$$

$$(\overline{BC})^2 + (\overline{AC})^2 = (\overline{AB})^2$$

Let $\overline{CD} = 50$, $\overline{BC} = 120$:

$$120^2 + 50^2 = x^2$$

$$14{,}400 + 2{,}500 = x^2$$

$$16{,}900 = x^2$$

Take the square root:

$$130 = x$$

Symmetric Property:

$$x = 130$$

Answer

130 feet

1a. A ladder touches a wall at a point 9 feet above the ground. If the ladder touches the ground at a point 12 feet away from the foot of the building, how long is the ladder?

1b. The sides of a rectangle have lengths of 16 and 30. Find the diagonal of the rectangle.

Exercises

Example 2

Vincente flies a kite with a 250-foot long string. If the kite is directly above a point on the ground 100 feet away from Vincente, find the elevation of the kite. Round to the nearest foot.

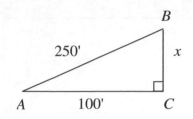

Analysis

Let x = the elevation of the kite.

Use the Pythagorean Theorem.

Work

Pythagorean
Theorem:

$$a^2 + b^2 = c^2$$

$$(\overline{BC})^2 + (\overline{AC})^2 = (\overline{AB})^2$$

$\overline{AC}$ = 100,
$\overline{AB}$ = 250:

$$x^2 + 100^2 = 250^2$$

$$x^2 + 10{,}000 = 62{,}500$$

Subtract 10,000:

$$x^2 = 52{,}500$$

Take the square
root:

$$x = 229.129 \approx 229$$

Answer

229 feet

Exercises

2a. An airplane takes off and ascends a diagonal distance of 5 miles. At that point it has traveled a distance of 4 horizontal miles. What is its altitude?

2b. A piece of property in the shape of a right triangle has one leg 80 feet long and a hypotenuse 170 feet long. What is the measure of the other leg?

Example 3

The length of a living room is 2 feet less than twice its width. If the diagonal is 2 feet more than twice the width, find the dimensions of the room.

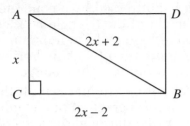

Analysis

Let x = the width of the living room.

Let $2x - 2$ = the length of the living room.

Let $2x + 2$ = the diagonal of the living room.

Use the Pythagorean Theorem.

Work

$$a^2 + b^2 = c^2$$
$$(\overline{BC})^2 + (\overline{AC})^2 = (\overline{AB})^2$$

$\overline{AC} = x,$
$\overline{BC} = 2x - 2,$
$\overline{AB} = 2x + 2:$

$$(2x - 2)^2 + x^2 = (2x + 2)^2$$
$$(4x^2 - 8x + 4) + x^2 = 4x^2 + 8x + 4$$
$$5x^2 - 8x + 4 = 4x^2 + 8x + 4$$

Subtract $4x^2$: $\qquad x^2 - 8x + 4 = 8x + 4$

Subtract $8x$: $\qquad x^2 - 16x + 4 = 4$

Subtract 4: $\qquad x^2 - 16x = 0$

Factor: $\qquad x(x - 16) = 0$

$$x = 0 \qquad x - 16 = 0$$
$$x = 16$$

Reject. $\qquad 2x - 2 = 30$

$$2x + 2 = 34$$

Check

$$x^2 + (2x - 2)^2 = (2x + 2)^2$$
$$16^2 + 30^2 = 34^2$$
$$256 + 900 = 1{,}156$$
$$1{,}156 = 1{,}156 \checkmark$$

Answer

Length = 30 feet, Width = 16 feet

3a.   A building is buttressed by a concrete right triangle. If the height of the triangle is 2 feet more than twice its horizontal base and the hypotenuse of the triangle is 3 feet more than twice its horizontal base, find the dimensions of the right triangle.

3b.   A horizontal railroad tunnel is built through the base of a hill of uniform grade. If the length of the tunnel is represented by x, the diagonal distance of the hill is represented by $2x - 2$ and the vertical height up the hill is represented by $x + 2$, find the length of the tunnel.

11.2 The Tangent Ratio

The right triangle is the most important triangle in trigonometry.

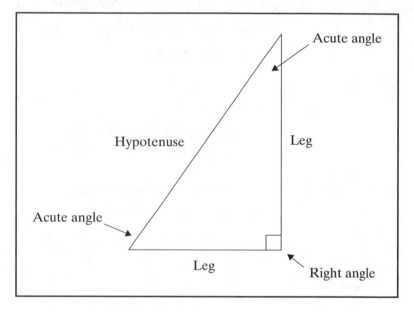

In a right triangle, if we compare the relationship of the side opposite an acute angle with the side adjacent to that angle, we have a ratio called the tangent (*tan* in abbreviated form).

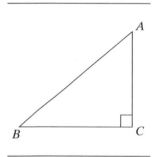

$$\tan \angle = \frac{\text{Opposite side}}{\text{Adjacent side}}$$

$$\tan A = \frac{\overline{BC}}{\overline{AC}} \qquad \tan B = \frac{\overline{AC}}{\overline{BC}}$$

Since ancient times mathematicians have noticed that each acute angle in a right triangle has a distinct tangent.

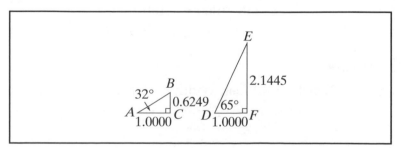

$$\tan 32° = \frac{0.6249}{1.0000} = 0.6249$$

$$\tan 65° = \frac{2.1445}{1.0000} = 2.1445$$

Now let's see what happens when we double the base of a triangle. Here the base increases from 1.0000 to 2.0000.

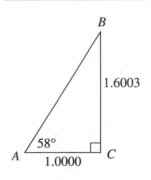

We have two similar triangles, so let's now use proportions in order to find $\overline{B'C'}$.

$$\frac{\overline{BC}}{\overline{AC}} = \frac{\overline{B'C'}}{\overline{A'C'}}$$

Given that $\overline{BC} = 0.6249$, $\overline{AC} = 1.0000$, $\overline{B'C'} = x$, and $\overline{A'C'} = 2.0000$:

$$\frac{0.6249}{1.0000} = \frac{x}{2.0000}$$

$$x = 1.2498$$

$\overline{B'C'}$, at 1.2498, is double $\overline{BC}$—just as $\overline{A'C'}$ is double $\overline{AC}$. The proportions have been preserved.

Angle	Tangent
16°	0.2867
35°	0.7002
58°	1.6003
70°	2.7475
77°	4.3315

Mathematicians have developed a table of values for tangents of the various acute angles in a right triangle. As an example, let's select one angle, say 58°, and see what this table tells us.

$$\tan 58° = \frac{1.6003}{1.0000} = 1.6003$$

This means that, the side opposite a 58° angle in a right triangle is 1.6003 times the length of the side adjacent to that angle.

We used to have to look up the values of trigonometric functions in tables; with the introduction of the scientific calculator, however, we can now skip the tables and go directly to the calculator.

Using a Scientific Calculator

For example, in the illustration above, in order to find the tangent of 58° with a calculator, just enter 58, then press the tan key.

$$\boxed{58} \quad \boxed{\tan} \quad = \quad \boxed{1.6003}$$

On the other hand, if we're given the tangent of an angle and we want to find the angle, we have to use the "inverse" key on the calculator. In the example above, just plug in 1.6003, then press the inverse key, followed by the tangent key.

$$\boxed{1.6003} \quad \boxed{\text{inv}} \quad \boxed{\tan} \quad = \quad \boxed{58}$$

Example 1

Marcia is standing 80 feet away from the base of a building. If the *angle of elevation* of the top of the building is 35°, find the height of of the building. Round to the nearest foot.

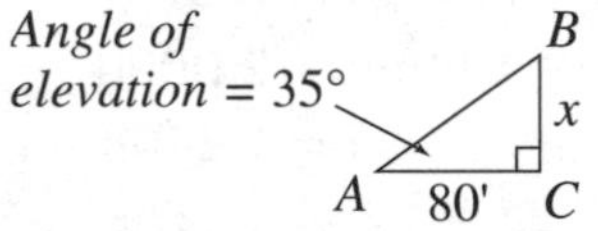

Analysis

Let x = the height of the building.

Use the tangent ratio.

$$\tan \angle = \frac{\text{Opposite side}}{\text{Adjacent side}}$$

Work

$$\tan 35° = \frac{x}{80}$$

$$0.7002 = \frac{x}{80}$$

$$x = 56.016 \approx 56$$

Answer

56 feet

Exercises

1a. A slide in a local park is built with an angle of elevation of 22°. If the end of the slide is 12 horizontal feet away from the start of the slide, what is the vertical height of the structure? Round the answer to the nearest hundredth of a foot.

1b. Nilda is in her third-floor apartment and is looking down at Karen. If the angle of depression is 29° and Nilda's apartment is 38 feet above the ground, how far away from the building is Karen standing? Round the answer to the nearest tenth of a foot.

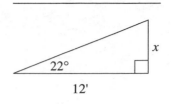

Example 2

Helene is on a weather balloon. She sights a lake 4,000 horizontal feet away. If her line of sight makes an angle of 56° with a perpendicular dropped from the balloon to the ground, how high is the balloon? Round the answer to the nearest foot.

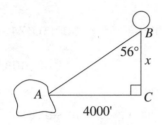

Analysis

Let m∠B = 56°.

Let $\overline{AC}$ = 4,000.

Let $\overline{BC}$ = x.

We have an opposite and an adjacent side, so we can use the tangent ratio.

Work

$$\tan \angle = \frac{\text{Opposite side}}{\text{Adjacent side}}$$

There are two methods of solving this problem. The first solution is to use the tangent of ∠B.

$$(1) \quad \tan 56° = \frac{4,000}{x}$$

tan 56° = 1.4826:

$$1.4826 = \frac{4,000}{x}$$

Multiply by x:

$$x\left(1.4826 = \frac{4,000}{x}\right)$$

$$1,4826x = 4,000$$

Divide by 1.4826:

$$x = 2,697.963 \approx 2,698$$

A second solution is to use the tangent of ∠A. However, to use this function, we need to find the measure of ∠A. We know that the sum of the angles of a triangle add up to 180°:

$$m\angle A + m\angle B + m\angle C = 180°$$

m∠C = 90°,
m∠B = 56°:

$$m\angle A + 56° + 90° = 180°$$

$$m\angle A + 146° = 180°$$

Subtract 146:

$$m\angle A = 34°$$

Now we'll use the tangent of 34°.

$$(2) \quad \tan 34° = \frac{x}{4,000}$$

tan 34° = 0.6745:

$$0.6745 = \frac{x}{4,000}$$

Multiply by 4,000:

$$4,000\left(0.6745 = \frac{x}{4,000}\right)$$

$$2,698 = x$$

Symmetric Property:

$$x = 2,698$$

Notice that we get the same answer using either method.

Answer

2,698 feet

Exercises

2a. A rectangle is 34 feet long. If the diagonal makes an angle of 72° with the width, find the width of the rectangle, to the nearest hundredth of a foot.

2b. The Leaning Tower of Pisa makes an angle of 85° with the ground. If the top of the Tower is 178 vertical feet above the ground, how much out of plumb is the Tower (side x)? Round to the nearest tenth of a foot.

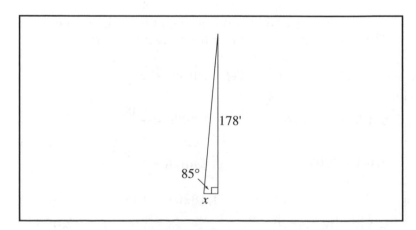

Example 3

A lighthouse is 200 feet high. From the top of the lighthouse find the *angle of depression*—to the nearest degree—of a boat out at sea if the boat is 3,000 feet away from the foot of the lighthouse.

Angle of depression

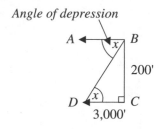

Analysis

Let x = the angle of depression.

Lines AB and CD are parallel, so their alternate interior angles are equal. The angle of elevation and depression are alternate interior angles and are therefore congruent: m$\angle BDC \cong$ m$\angle x$.

Work

$$\tan\angle BDC = \frac{\text{Opposite side}}{\text{Adjacent side}} = \frac{200}{3,000} = 0.0667$$

We're now looking for an angle whose tangent is 0.0067, so we'll have to go to the Table of Trigonometric Functions, or else use the inverse trigonometric function on the scientific calculator. The angle whose tangent is 0.0667 is located in between 3° and 4°.

Angle	Tangent			
			0.0667	0.0699
			−0.0524	−0.0667
3°	0.0524	} 0.0143	0.0143	0.0032
x	0.0667			
4°	0.0699	} 0.0032		

We can see that 0.0667 is closer to 0.0699 than it is to 0.0524.

Answer

m$\angle x = 4°$

3a. The Empire State Building is 1,250 feet tall. To the nearest degree, find the angle it subtends at a distance of 942 feet away from the base.

3b. A diagonal street runs 942 feet west and 420 feet north. To the nearest degree, find the angle it makes with a vertical line.

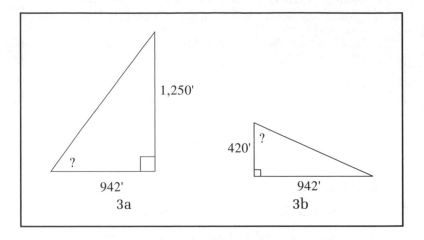

3a   3b

11.3 The Sine Ratio

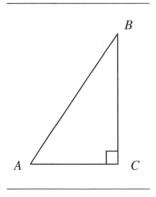

In a right triangle, if we compare the side opposite an acute angle with the hypotenuse, we have a ratio called the sine (*sin*, in abbreviated form).

$$\sin\angle = \frac{\text{Opposite side}}{\text{Hypotenuse}}$$

$$\sin\angle A = \frac{\overline{BC}}{\overline{AB}} \qquad \sin\angle B = \frac{\overline{AC}}{\overline{AB}}$$

Example 1

A road is inclined 8° to the horizontal. Determine to the nearest tenth of a foot the rise for each 500 feet of roadway.

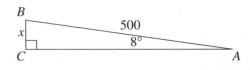

Analysis

Let $\overline{AB}$ = the roadway.

Let x = the rise of the roadway.

Let m∠A = 8°.

We have a side opposite ∠A and the hypotenuse, $\overline{AB}$, so we can use the sine function:

$$\sin \angle A = \frac{\text{Opposite side}}{\text{Hypotenuse}}$$

$$\sin 8° = \frac{x}{500}$$

Sin 8° = 0.1392:

$$0.1392 = \frac{x}{500}$$

Multiply by 500:

$$500\left(0.1392 = \frac{x}{500}\right)$$

$$69.6 = x$$

Symmetric Property:

$$x = 69.6$$

Answer

69.6 feet

1a. An airplane takes off at an angle of 9° with the runway. What is the plane's altitude after flying a diagonal distance of 3,400 feet? Round to the nearest foot.

1b. A wire from the top of the Eiffel Tower to a point in the ground makes a 23° angle with the ground. If the wire is 2,419 feet long, what is the height of the Eiffel Tower? Round to the nearest foot.

Exercises

Example 2

A 90-foot wire is attached to the top of a pole and is anchored to the ground some distance away. If the wire makes an angle of 19° with the ground, how far away from the foot of the pole is the wire anchored to the ground? Round off the answer to the nearest tenth of a foot.

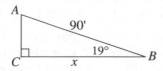

Analysis

Let $x = \overline{BC}$ = distance from the pole.

Let $\overline{AB} = 90'$.

Let $\angle B = 19°$.

To use the sine function, we need to use an opposite side and the hypotenuse. We don't know what $\overline{AC}$—the side opposite $\angle B$—is, so we need to use another approach. Let's find m$\angle A$. After we find $m\angle A$, we have an opposite side—$\overline{BC}$, or x—and we have the hypotenuse, $\overline{AB}$.

Work

The sum of the angles of a triangle is 180°:

$$m\angle A + m\angle B + m\angle C = 180°$$

$$m\angle A + 19° + 90° = 180°$$

$$m\angle A + 109° = 180°$$

Subtract 109: $\qquad\qquad\qquad m\angle A = 71°$

Now we can use
$\sin \angle A$: $\qquad\qquad \sin\angle A = \dfrac{\text{Opposite side}}{\text{Hypotenuse}}$

$$\sin 71° = \frac{x}{90}$$

Sin 71° = 0.9455: $\qquad\qquad 0.9455 = \dfrac{x}{90}$

Multiply by 90: $\qquad\qquad 90\left(0.9455 = \dfrac{x}{90}\right)$

$$85.095 = x$$

Symmetric Property: $\qquad\qquad x = 85.095 \approx 85.1$

Answer

85.1 feet

Exercises

2a. Firefighters are rescuing some people caught in a flaming apartment. The firefighters are on a ladder 90 feet long that is leaning against a windowsill. If the ladder makes an angle of 87° with the ground, how far away from the base of the building is the ladder on the ground? Round to the nearest tenth of a foot.

2b. The diagonal of a rectangle is 600 feet. If the diagonal makes an angle of 54° with a side of the rectangle, find that side. Round to the nearest foot.

Example 3

An airplane takes off at an angle of 10° with the ground. Find the distance the plane has flown when it reaches an altitude of 4,500 feet. Round to the nearest foot.

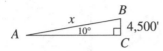

Analysis

Let x = the distance the plane has flown.

Let $\overline{BC}$ = 4,500′. Use the sine function.

Work

$$\sin \angle = \frac{\text{Opposite side}}{\text{Hypotenuse}}$$

$$\sin 10° = \frac{4,500}{x}$$

Sin 10° = 0.1736:
$$0.1736 = \frac{4,500}{x}$$

Multiply by x:
$$x\left(0.1736 = \frac{4,500}{x}\right)$$

$$0.1736x = 4,500$$

Divide by 0.1736:
$$x = 25,921.6 \approx 25,922$$

Answer

25,922 feet

Exercises

3a.   A car moves up a mountain road of uniform grade. If the road makes a 7° angle with the horizontal and the car attains a vertical height of 2,700 feet, how far did the car travel along the road? Round to the nearest foot.

3b.   A pole is 34 feet tall. A wire attached to the pole at one end and to the ground at the other end makes an angle of 58° with the ground. To the nearest tenth of a foot, how long is the wire?

Example 4

A rafter for a roof is 15 feet long and the rise is 7 feet. To the nearest degree, find the *angle of elevation*.

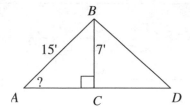

Analysis

Let $\overline{BC} = 7$.

Let $\overline{AB} = 15$.

Let $\angle A$ = the angle of elevation.

We know the side opposite $\angle A$ and the hypotenuse, so we can use the sine function.

Work

$$\sin\angle A = \frac{\text{Opposite side}}{\text{Hypotenuse}} = \frac{7}{15} = 0.4667$$

Check the Table of Trigonometric Functions, or else use the inverse trigonometric function on the scientific calculator. In the table, look under the sine values and locate 0.4667 between 0.4540 and 0.4695.

Angle	Sine
27°	0.4540
	0.4667
28°	0.4695

0.4540 ⎱ 0.0127
0.4667 ⎰
0.4695 ⎱ 0.0028

```
  0.4667      0.4695
 -0.4540     -0.4667
 _____     _____
  0.0127      0.0028
```

Because 0.4667 is closer to 0.4695 than to 0.4540, to the nearest degree, m$\angle A = 28°$.

Answer

28°

4a. The height of a right triangle is 14 feet. If its hypotenuse is 19 feet, to the nearest degree, what is the acute angle at the base of the triangle?

4b. A railroad track ascends a mountain of uniform grade. For every 600 feet of track it moves a horizontal distance of 325 feet. Find the angle the track makes with a vertical line. Round to the nearest degree.

Example 5

An 11-foot ladder leans against a wall. If the ladder touches a point on the wall 6 feet above the ground, find the angle formed by the ladder and the wall.

Analysis

Let $\angle B$ = the angle formed by the ladder and the wall.

Let $\overline{BC}$ = 6.

$m\angle C = 90°$.

We will use the sine function, so we need to know the measure of the side opposite acute angle B (side $\overline{AC}$) and the measure of the hypotenuse (side $\overline{AB}$). We know the measure of the hypotenuse, but we don't know the measure of $\overline{AC}$, so we'll try to use another approach.

We know the side opposite $\angle A$ (side $\overline{BC}$) and we know the hypotenuse ($\overline{AB}$), so let's use the sin of $\angle A$ to find its measure. Then, since the measures of the angles in a triangle add up to 180°, we'll just substitute and find $m\angle B$.

Work

$$\sin\angle A = \frac{\text{Opposite side}}{\text{Hypotenuse}} = \frac{\overline{BC}}{\overline{AB}} = \frac{6}{11} = 0.5455$$

Locate $\angle A$ between 33° and 34°, or else use the inverse trigonometric function on the scientific calculator.

Angle	Sine		
		0.5455	0.5592
		−0.5446	−0.5455
33°	0.5446 ⎤ 0.0009	0.0009	0.0137
A	0.5455 ⎦		
34°	0.5592 ⎦ 0.0137		

To the nearest degree, m∠A = 33°.

We know that the sum of the angles of a triangle equal 180°, so we'll just substitute 33° for m∠A in the following equation:

$$\text{m}\angle A + \text{m}\angle B + \text{m}\angle C = 180°$$

$$33 + \text{m}\angle B + 90° = 180°$$

$$123 + \text{m}\angle B = 180°$$

$$\text{m}\angle B = 57°$$

Answer

$$\text{m}\angle B = 57°$$

Exercises

5a. An airplane ascends at an angle for a diagonal distance of 5,500 feet. At that moment it has attained a vertical height of 4,700 feet. To the nearest degree, find the angle the diagonal flight of the plane makes with a vertical line to the ground.

5b. A batter hits a baseball a distance of 500 feet until it crashes into the outfield wall. At that point, the ball has traveled a horizontal distance of 460 feet. To the nearest degree, find the angle the line of flight of the ball makes with the ground.

11.4 The Cosine Ratio

In a right triangle, if we compare the side adjacent to an acute angle with the hypotenuse, we have a ratio called the cosine (*cos*, in abbreviated form).

$$\cos\angle = \frac{\text{Adjacent side}}{\text{Hypotenuse}}$$

$$\cos\angle A = \frac{\overline{AC}}{\overline{AB}} \qquad \cos\angle B = \frac{\overline{BC}}{\overline{AB}}$$

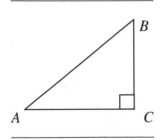

Example 1

Julio flies a kite. He lets out 250 feet of string, and the string makes an angle of 32° with the ground. What is Julio's *horizontal distance* from the kite? Round the answer to the nearest foot.

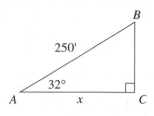

Analysis

Let m∠ = 32°

Let string $\overline{AB}$ = 250'.

Let $\overline{AC}$ = x.

We have an adjacent side and the hypotenuse, so we'll use the cosine ratio.

Work

$$\cos\angle = \frac{\text{Adjacent side}}{\text{Hypotenuse}}$$

$$\cos 32° = \frac{x}{250}$$

Cos 32° = 0.8480: $\qquad 0.8480 = \dfrac{x}{250}$

Multiply by 250: $\qquad 250\left(0.8480 = \dfrac{x}{250}\right)$

$$212 = x$$

Symmetric Property: $\qquad x = 212$

Answer

212 feet

1a. The hypotenuse of a right triangle is 18 feet. To the nearest tenth of a foot, find the base if the acute angle at the base is 15°.

1b. A car travels up a mountain road of a uniform rise 4,800 feet. If the angle the road makes with a vertical line is 49°, find the vertical rise of the road. Round to the nearest foot.

Exercises

Example 2

The diagonal of a rectangular parcel of land makes an angle of 18° with the longer side. If the longer side is 120 feet, find the length of the diagonal. Round to the nearest tenth of a foot.

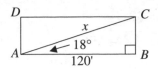

Analysis

Let the diagonal = x.

We have an adjacent side and the hypotenuse, so we'll use the cosine function.

Work

$$\cos \angle BAC = \frac{\text{Adjacent side}}{\text{Hypotenuse}}$$

$$\cos 18° = \frac{120}{x}$$

cos 18° = 0.9511: $0.9511 = \dfrac{120}{x}$

Multiply by x: $x\left(0.9511 = \dfrac{120}{x}\right)$

$$0.9511x = 120$$

Divide by 0.9511: $x = 126.170 \approx 126.2$

Answer

126.2 feet

Exercises

2a.   A ladder leans against a wall and makes an angle of 61° with the ground. If the base of the ladder touches a point 5 feet from the wall, how long is the ladder? Round to the nearest tenth of a foot.

2b.   In right triangle ABC, m$\angle C = 90°$, $\overline{BC} = 18$, and m$\angle B = 75°$. Find the length of the hypotenuse. Round to the nearest tenth.

Example 3

A portion of an inclined road is 5,500 feet. Find the angle of elevation of this portion of the road if it covers a horizontal distance of 4,000 feet.

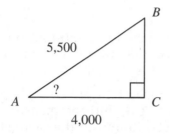

Analysis

Let $\angle A$ = the angle of elevation.

Let $\overline{AC}$ = 4,000.

Let $\overline{AB}$ = 5,500.

We have an adjacent side and the hypotenuse, so we'll use the cosine function.

Work

$$\cos\angle = \frac{\text{Adjacent side}}{\text{Hypotenuse}}$$

$$\cos\angle A = \frac{4,000}{5,500} = 0.7273$$

Locate $\angle A$ between 43° and 44°, or else use the inverse trigonometric function on the scientific calculator.

Angle	Cosine	
43°	0.7314	⎫ 0.0041
x	0.7273	⎬
44°	0.7193	⎭ 0.0080

0.7314	0.7273
−0.7273	−0.7193
0.0041	0.0080

Because 0.7273 is closer to 0.7314 than it is to 0.7193, to the nearest degree, $m\angle A = 43°$.

Answer

43°

3a. A railroad track runs up an inclined mountain road of uniform grade. If the track runs 1,200 feet for every 980 feet of horizontal distance, find the angle of elevation of the track.

3b. In rectangle $ABCD$, diagonal $\overline{BD} = 200$ and side $\overline{BC} = 72$. Find m$\angle CBD$, to the nearest degree.

Test

1–4. @ 5%

1. A cable is attached to A 30-foot pole. If the cable touches a point on the ground 16 feet from the base of the pole, how long is the cable?

2. The hypotenuse of a right triangle is 16 feet. If the acute base angle is 19°, find the measure of the side opposite this angle. Round to the nearest tenth of a foot.

3. An engineer sets up his transit (5 feet tall) 90 feet from the base of building. If his angle of elevation at the top of the building is 51°, how tall is the building? Round to the nearest tenth of a foot, and add the height of the transit to your result.

4. A ladder leans against a wall and rests on the ground at a point 12 feet from the base of the wall. If the ladder is 15 feet long, how far from up from the ground does it touch the wall?

5–14. @ 8%

5. A road is inclined 8°. If a car ascends to a height of 1,400 feet, how much of a horizontal distance has it traveled? Round to the nearest foot.

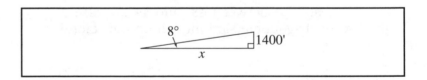

6. Jenny goes up to the top of the Sears Tower and spots a bus in the street below. If the Sears Tower is 1,454 feet tall and the bus is 3,425 feet away from the base of the tower, what is the angle formed by Jenny's line of sight and the Sears Tower? Round the answer to the nearest whole degree.

7. A horizontal tunnel is built through the base of a mountain of uniform grade. If the length of the tunnel is represented by $x - 2$, the diagonal distance to the top of the mountain is represented by $x + 2$, and the vertical height of the mountain is represented by x, find the length of the tunnel.

8.   A road runs up the side of a mountain of uniform grade. If the elevation increases 97 feet for every 150 feet of road, to the nearest degree, what is the angle of elevation nearest the road?

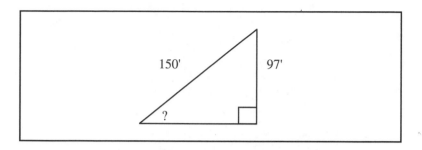

9.   North-south street $\overline{BC}$ intersects east-west street $\overline{AC}$ at right angle C. Street $\overline{AB}$ is diagonal and cuts across streets $\overline{BC}$ and $\overline{AC}$. If $\overline{AB}$ is 334 feet and m$\angle BAC = 61°$, find the length of street $\overline{AC}$. Round the answer to the nearest foot.

10.   Michael is flying a kite. If the kite is 300 feet directly above the ground and the string makes an angle of 48° with an imaginary vertical line, how long is the string? Round the answer to the nearest foot.

11.   In an emergency, airline personnel attach slides for passengers for quick exits. If the base of the slide touches the ground at a point 25 feet away from the airplane and the slide makes an angle of 23° with the ground, how high above the ground is the slide anchored to the emergency exit of the plane? Round your answer to the nearest tenth of a foot.

12.   A missile is projected upward at a 37° angle to the horizontal. If the missile covers a diagonal distance of 3,900 feet, what is its altitude above the ground? Round to the nearest foot.

13.   Doug cuts across the diagonal of a rectangular lot. If the lot is 60 feet long and his diagonal path is 82 feet, what is the angle formed by the diagonal and the long side of the lot?

14.   The hypotenuse of a right triangle is 26. If the base is 10, find the altitude.

12

Trigonometry of General Triangles

The very name *trigonometry* is derived from the ancient Greek and it literally means the measurement of *trigons*, the Greek word for triangles. Trigonometry has its roots in antiquity. The early Egyptians made extensive use of trigonometry in surveying and in constructing pyramids. The ancient Greeks were instrumental in adapting trigonometric methods for use in astronomy. And, in order to sail the oceans, ancient navigators had to know some geometry and trigonometry. They had to calculate their own position in relation to the North Pole and certain markers in the constellations.

In the third century B.C.E., Greek mathematicians began recording the lengths of chords ($\overline{AB}$ and $\overline{AC}$) subtending a sequence of larger and larger arcs in a circle of unit radius 1. In essence, this was the beginning of the sine table. Two hundred years later, trig tables were developed for chords whose arcs increased in multiples of 7.5°.

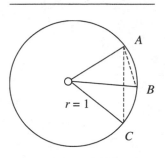

In the second century C.E., the Greek scientist Ptolemy included in his *Almagest* a table of such chords in increments of 0.25°.

We often compare triangles that are exact duplicates of one another. These triangles are called **congruent (≅) triangles**.

12.1 Similar Triangles

Triangles that are not congruent but whose sides are in proportion are also important and interesting. These are called **similar (~) triangles**. In these cases, we have to know which sides and angles are corresponding to set up a proportion.

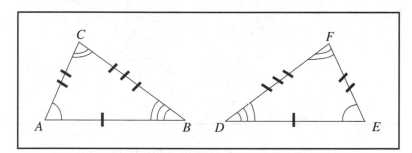

In triangles *ACB* and *EFD*, the following angles are congruent: $\angle A \cong \angle E$, $\angle B \cong \angle D$, $\angle C \cong \angle F$. The following sides are congruent: $\overline{AB} \cong \overline{DE}$, $\overline{AC} \cong \overline{EF}$, $\overline{BC} \cong \overline{DF}$.

We locate corresponding sides opposite congruent angles. Congruent angles are located opposite corresponding sides.

Congruent Angles	Corresponding Sides
$\angle B \cong \angle D$	$\overline{AC}$ and $\overline{EF}$
$\angle C \cong \angle F$	$\overline{AB}$ and $\overline{DE}$
$\angle A \cong \angle E$	$\overline{BC}$ and $\overline{DF}$

What exactly do we mean when we talk about similar triangles? By definition, similar triangles fulfill the following conditions:

1. Their corresponding angles are congruent.

2. Their corresponding sides are in proportion.

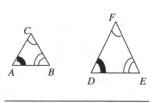

According to the definition of similar triangles, if we are given that $\angle A \cong \angle D$, $\angle B \cong \angle E$, and $\angle C \cong \angle F$, we can draw the following conclusions:

1. $\triangle ABC$ is similar to (~) $\triangle DEF$.

2. The corresponding sides are in proportion:

$$\frac{\overline{AB}}{\overline{DE}} = \frac{\overline{AC}}{\overline{DF}} = \frac{\overline{BC}}{\overline{EF}}$$

To determine whether given triangles are similar, we don't always need to fulfill the complete definition of similar triangles—that the corresponding angles are congruent and the corresponding sides are in proportion. There are shortcuts in determining whether triangles are similar.

Let's look at triangles ABC and $A'B'C'$. We're given the following information:

$$\angle A \cong \angle A'$$

$$\angle B \cong \angle B'$$

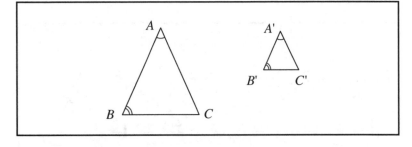

If two angles of one triangle are congruent to two corresponding angles in a second triangle, the third angles in both triangles are also congruent. We can now assert that $\angle C \cong \angle C'$

Let's move $\triangle A'B'C'$ and place it over $\triangle ABC$ so that $\angle A'$ coincides with $\angle A$. Since $\angle B \cong \angle B'$ $\overline{BC} \parallel \overline{B'C'}$. Now, by a Euclidean proposition (which can be proven), if a line is parallel to one side of a triangle, it divides the other two sides proportionally. We can therefore conclude that

$$\frac{\overline{A'B'}}{\overline{AB}} = \frac{\overline{A'C'}}{\overline{AC}}$$

Let's repeat the procedure except now we'll place $\triangle A'B'C'$ over $\triangle ABC$ so that $\angle C'$ coincides with $\angle C$. This time we derive the proportion

$$\frac{\overline{A'C'}}{\overline{AC}} = \frac{\overline{B'C'}}{\overline{BC}}$$

By the transitive property of equality,

$$\frac{\overline{A'B'}}{\overline{AB}} = \frac{\overline{A'C'}}{\overline{AC}} = \frac{\overline{B'C'}}{\overline{BC}}$$

Shortcut 1 in Proving Triangles Congruent

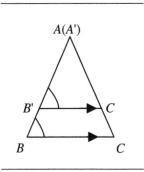

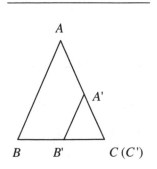

Do three angles of one triangle need to be congruent (≅) to three angles of a second triangle for the corresponding sides to be in proportion and to have similar triangles? No, since the three angles add up to 180°, if we know two of the angles, we just have to subtract from 180° to find the third angle. Therefore, it's sufficient to know that two angles of one triangle are congruent to two angles of a second triangle in order for the two triangles to be similar.

Conclusion

It's sufficient to show that, if we have a minimum of two angles of one triangle congruent to two angles of a second triangle, the corresponding sides are in proportion and we have two similar triangles.

Now, let's investigate the case whereby we have two triangles, all of whose sides are in proportion:

$$\frac{\overline{A'B'}}{\overline{AB}} = \frac{\overline{A'C'}}{\overline{AC}} = \frac{\overline{B'C'}}{\overline{BC}}$$

By a Euclidean proposition (which can be proven), if a line divides two sides of a triangle proportionally, it is parallel to the third side.

Now that we have shown that $\overline{B'C'}$ is parallel to $\overline{BC}$, the corresponding angles are congruent:

$$\angle B \cong \angle A'B'C' \qquad \angle C \cong \angle A'C'B'$$

If two angles of one triangle are congruent to two angles of a second triangle, the third angles are also congruent ($\angle A \cong \angle A'$).

Conclusion

It's sufficient to show that, if we have the corresponding sides of one triangle in proportion of the corresponding sides of a second triangle, the corresponding angles are congruent, and we have two similar triangles.

Shortcut 2 in Proving Triangles Congruent

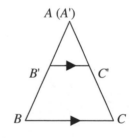

Example 1

A 40-foot flagpole casts a 25-foot shadow. Find the shadow cast by a nearby building 200 feet tall.

Analysis

$\angle C \cong \angle F$ because they're both right triangles. $\angle B \cong \angle E$ because the sun's angle of elevation is the

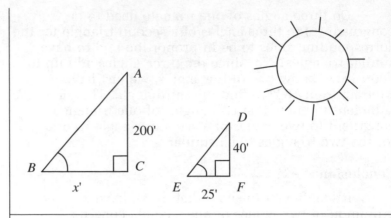

same. Now, we have two angles of one triangle congruent to two angles of a second triangle. We have shown that it is sufficient to have two similar triangles. Once we have two similar triangles, we can set up a proportion to solve for one missing side.

Work

$$\frac{\overline{AC}}{\overline{BC}} = \frac{\overline{DF}}{\overline{EF}}$$

$\overline{AC} = 200, \overline{BC} = x, \overline{DF} = 40, \overline{EF} = 25$:

$$\frac{200}{x} = \frac{40}{25}$$

$$40x = 5{,}000$$

Divide by 40:

$$x = 125$$

Answer

125 feet

Exercises

1a.   In $\triangle ABC$, m$\angle A = 25°$, m$\angle B = 74°$, $\overline{AC} = 12$, and $\overline{AB} = 16$. In $\triangle DEF$, m$\angle D = 25°$, m$\angle E = 74°$, and $\overline{DE} = 30$. Find DF.

1b.   An automobile headlight is installed so that the angle between its beam of light and an imaginary line from the ground up through its center is 88°. If the headlight is installed 4 feet above the ground, the beam hits the ground at a horizontal distance of 114.5 feet. Find the distance the beam will hit the ground if the headlight is installed 3.8 feet above the ground. Round to the nearest tenth of a foot.

Example 2

Find a way for Jesus to measure the width of a river, $\overline{AC}$, without actually going into the water or crossing the river.

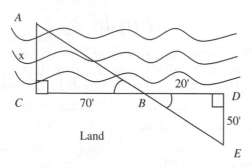

Analysis

From point C on land, Jesus measures off 70 feet and then sets up a stake at point B. He then measures 20 feet from point B to point D. At this point, he makes a right angle turn and walks 50 feet to point E.

The distance across the river, $\overline{AC}$, is perpendicular to $\overline{CD}$. $\angle C \cong \angle D = 90°$. $\angle ABC \cong \angle DBE$ because all vertical angles are congruent. The sides opposite congruent angles are corresponding sides. $\overline{BE}$ (opposite right $\angle D$) and $\overline{AB}$ (opposite right $\angle C$) are corresponding sides. $\overline{DE}$ (opposite $\angle DBE$) and $\overline{AC}$ (opposite $\angle ABC$) are also corresponding sides.

Jesus can now set up two similar triangles with corresponding sides in proportion to determine the width of the river, $\overline{AC}$.

Work

$$\frac{\overline{AC}}{\overline{BC}} = \frac{\overline{DE}}{\overline{DB}}$$

$\overline{AC} = x$, $\overline{BC} = 70$, $\overline{DE} = 50$, $\overline{DB} = 20$:

$$\frac{x}{70} = \frac{50}{20}$$

$$20x = 3{,}500$$

Divide by 20:

$$x = 175$$

Answer

175 feet

2a. Lines $\overline{AB}$ and $\overline{CD}$ intersect at E. If $\overline{DE} = 15$, $\overline{BE} = 35$, $\overline{AE} = 12$, m$\angle D = 23°$, and m$\angle B = 23°$, find $\overline{CE}$.

2b. A 110-foot building casts a shadow of 88 feet. How long is the shadow cast by a nearby 40-foot tall flagpole?

Example 3

The Akron Architectural Firm is making a model of a one-family house. At this time, the firm is developing a model for the roof. In the model, $\overline{AC} = 11''$, $\overline{AB} = 12''$, $\overline{BC} = 5.3''$, m$\angle A = 26°$, and m$\angle B = 24°$. In the actual roof, $\overline{DF} = 99''$, $\overline{DE} = 108''$, and $\overline{BC} = 47.7''$. Find the measure of $\angle F$ in the actual roof.

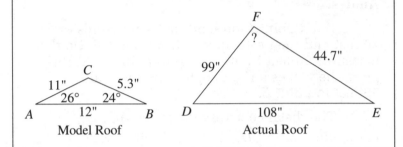

Model Roof   Actual Roof

Analysis

Let's first determine whether the two triangles are similar.

One way to determine whether two triangles are similar is to have three angles in one triangle congruent to three angles in the second triangle. We don't have that situation here.

A second method of determining similarity is see whether three sides in one triangle are in proportion to three sides in a second triangle. Let's check for this situation:

$$\frac{\overline{AC}}{\overline{DF}} \; ? \; \frac{\overline{AB}}{\overline{DE}} \; ? \; \frac{\overline{CB}}{\overline{FE}}$$

$$\frac{11}{99} = \frac{12}{108} = \frac{5.3}{47.7}$$

$$\frac{1}{9} = \frac{1}{9} = \frac{1}{9}$$

The sides are in proportion, so now we can say that the corresponding angles are congruent:

$$\angle A \cong \angle D$$

$$\angle B \cong \angle E$$

$$\angle C \cong \angle F$$

Now, let's find m$\angle C$ in $\triangle ABC$. The sum of the angles in a triangle add up to 180°:

$$m\angle A + m\angle B + m\angle C = 180°$$

$m\angle A = 26°, m\angle B = 24°$: $26° + 24° + m\angle C = 180°$

$$50° + m\angle C = 180°$$

Subtract 50: $m\angle C = 130°$

$\angle C$ and $\angle F$ are corresponding angles in similar triangles and are, therefore, congruent, so m$\angle C$ = m$\angle F$ = 130°.

Answer

130°

3a.   In $\triangle RST$, $\overline{RS}$ = 8″, $\overline{RT}$ = 4″, $\overline{ST}$ = 6″, m$\angle R$ = 47°, and m$\angle T$ = 15°. In $\triangle KLM$, $\overline{KL}$ = 20″, $\overline{LM}$ = 15″, and $\overline{KM}$ = 12″. Are the two triangles similar? If they are, find m$\angle L$ in $\triangle KLM$.

Exercises

3b.   Diamond Aircraft is designing a new triangular airplane wing. The model is on the left side and the actual wing is on the right side below.

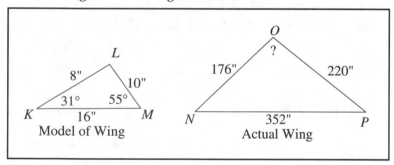

Model of Wing   Actual Wing

$\overline{KL}$ = 8″, $\overline{LM}$ = 10″, $\overline{KM}$ = 16″, $\overline{NP}$ = 352″, $\overline{NO}$ = 176″, $\overline{PO}$ = 220″, m$\angle K$ = 31°, and m$\angle M$ = 55°. Are the two triangles similar? If they are, find the measure of $\angle O$.

Example 4

In the diagram $\overline{BD}$ is parallel to $\overline{AE}$. $\overline{BC}$ = x, $\overline{BA}$ = x + 2, $\overline{CD}$ = x + 1, $\overline{DE}$ = x + 4. Find $\overline{BC}$.

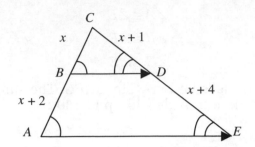

Analysis

We have two overlapping triangles, $\triangle ACE$ and $\triangle BCD$.

Since $\overline{BD}$ is parallel to $\overline{AE}$, their corresponding angles are congruent:

$$\angle CBD \cong \angle CAE$$

$$\angle CDB \cong \angle CEA$$

If we have two angles of one triangle congruent to two angles of a second triangle, the triangles are similar. We can now set up a proportion between the corresponding sides of both triangles.

Work

$$\frac{\overline{AC}}{\overline{BC}} = \frac{\overline{EC}}{\overline{DC}}$$

$\overline{AC} = (x+2) + x = 2x+2$, $\overline{BC} = x$,
$\overline{EC} = (x+1) + (x+4) = 2x+5$, $\overline{DC} = x+1$:

$$\frac{2x+2}{x} = \frac{2x+5}{x+1}$$

Multiply by $x(x+1)$:
$$x(x+1)\left(\frac{2x+2}{x} = \frac{2x+5}{x+1}\right)$$

$$(x+1)(2x+2) = x(2x+5)$$

$$2x^2 + 4x + 2 = 2x^2 + 5x$$

Subtract $2x^2$:
$$4x + 2 = 5x$$

Subtract $4x$:
$$2 = x$$

Symmetric Property:
$$x = 2$$

$$\overline{BC} = 2$$

Answer

2

4a. In the diagram below, $\overline{BD}$ is parallel to $\overline{AE}$, $\overline{AB} = 3x + 1$, $\overline{BC} = x + 3$, $\overline{BD} = x$, and $\overline{AE} = 3x$. Find $\overline{BC}$.

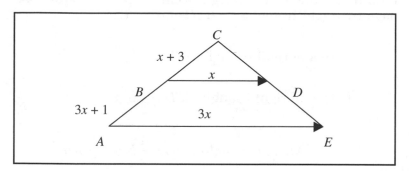

4b. $\overline{AD}$ and $\overline{BE}$ intersect at C. $\overline{AB}$ is parallel to $\overline{DE}$ so that $\angle B \cong \angle E$. Vertical angles 1 and 2 are congruent. If $\overline{DE} = x$, $\overline{AB} = 2x + 4$, $\overline{CD} = 2x + 2$, and $\overline{AC} = 8x - 2$, find $\overline{CD}$.

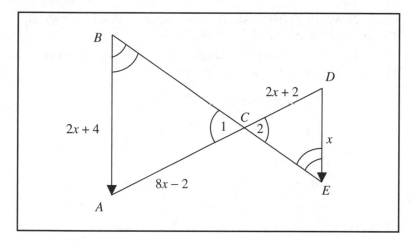

The area of a right triangle equals half the base times the height:

$$A = \frac{1}{2}bh$$

where b = base and h = height.

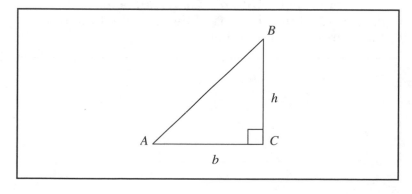

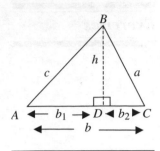

Let's consider the general triangle $\triangle ABC$, and we'll try to develop a formula for its area. First, drop a perpendicular, h, from B to side $\overline{AC}$. We know the formula for the area of a right triangle. At this point, we have two right triangles, $\triangle ABD$ and $\triangle BCD$.

$$\text{Area of right triangle } ABD = \frac{1}{2}b_1 h$$

$$+ \text{ Area of right triangle } BCD = \frac{1}{2}b_2 h$$

$$\text{Area of triangle } ABC = \frac{1}{2}b_1 h + \frac{1}{2}b_2 h$$

Factor:
$$= \frac{1}{2}h(b_1 + b_2)$$

Substitute $b = b_1 + b_2$:
$$= \frac{1}{2}hb$$

Now, in $\triangle ABD$, $\sin A = h/c$ or, $h = c \cdot \sin A$, where h = the height of $\triangle ABC$.

Let's now return to the equation for the area of triangle ABC and substitute $c \cdot \sin A$ for h:

$$\text{Area of triangle } ABC = \frac{1}{2}hb$$

$$= \frac{1}{2}c \cdot \sin A \cdot b$$

or:
$$= \frac{1}{2}bc \sin A$$

Conclusion

The area of any triangle equals half the product of any two sides and the sine of the included angle.

Example 1

In triangle ABC, $a = 29$, $c = 16$, and $m\angle B = 52°$. Find the area of the triangle. Round to the nearest whole number.

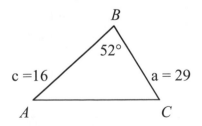

Analysis

We have two adjacent sides as well as the included angle.

Work

$$\text{Area} = \frac{1}{2}ac\sin B = 0.5(29)(16)\sin 52°$$
$$= 232(0.7880) = 182.82 \approx 183$$

Answer

183

1a.   In triangle ABC, $\overline{AB} = 22$, $\overline{BC} = 46$, and $m\angle ABC = 73°$. Find the area of the triangle and round to the nearest whole number.

1b.   The manager of the Real Right Shopping Center is going to purchase some turf to cover a triangular lawn. If two adjoining sides of the lawn are 123 feet and 187 feet and the included angle is 84°, what is the area of the lawn? Round to the nearest whole number.

Exercises

Example 2

One side of a roof is in the shape of a triangle. If two adjacent sides are 28 feet and 31 feet and the area of the triangle is 361 square feet, find the included angle. Round to the nearest degree.

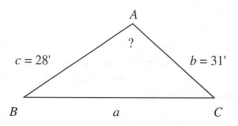

Analysis

We have two adjacent sides and the area. Use the area formula and then find the angle between the two adjacent sides.

Work

$$\text{Area} = \frac{1}{2}bc\sin A$$

$b = 31$, $c = 28$, Area = 361:

$$361 = 0.5(31)(28)\sin A$$
$$361 = 434\sin A$$
$$\sin A = 0.8318$$

Locate 0.8328 between 0.8290 and 0.8387, the sines of 56° and 58°, respectively, or else use the inverse trigonometric function on the scientific calculator.

0.8318	0.8387
−0.8290	−0.8318
0.0028	0.0069

Angle	Sine	
56°	0.8290	0.0028
	0.8318	0.0069
57°	0.8387	

The sine of the unknown angle is closer to the sine of 56°.

Answer

56°

Exercises

2a. In triangle ABC, $\overline{AC} = 80$, $\overline{BC} = 90$, and the area is 2,065. Find m∠C.

2b. The area of a triangular sail is 132 square feet. If two adjoining sides are each 19 feet, find the measure of the angle between these two adjoining sides.

12.3 Law of Cosines

Whereas the Pythagorean Theorem is specifically applicable for right triangles, we can derive a formula for the relationship of the sides in any triangle at all, called the Law of Cosines.

Angle A is an acute angle, so we'll locate it in quadrant 1.

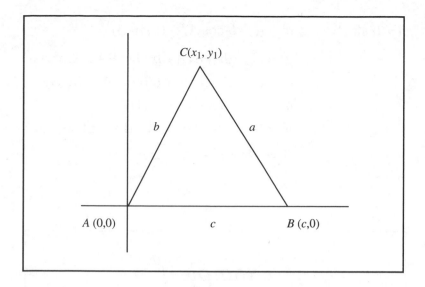

Let's drop a perpendicular from C to $\overline{AB}$, at point D.

Let the height $= h$.

Let $\overline{AD} = x_1$.

$\sin A = \dfrac{h}{b}$, so $h = b \cdot \sin A$

$\cos A = \dfrac{x_1}{b}$, so $x_1 = b \cdot \cos A$

Now $\overline{BD} = c - b \cdot \cos A$.

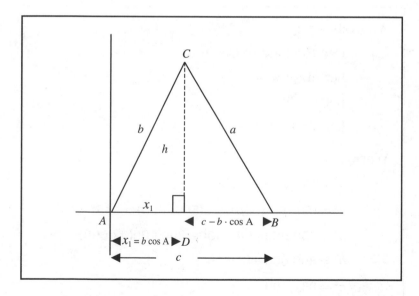

Let's use the Pythagorean Theorem and apply it to right triangle, $\triangle BDC$.

$$(\overline{BC})^2 = (\overline{BD})^2 + (\overline{CD})^2$$
$$a^2 = (c - b \cdot \cos A)^2 + h^2$$

$$h = b \sin A: \qquad a^2 = (c - b \cdot \cos A)^2 + (b \sin A)^2$$

$$a^2 = c^2 - 2bc \cos A + b^2 \cos^2 A + b^2 \sin^2 A$$

$$a^2 = c^2 - 2bc \cos A + b^2 (\cos^2 A + \sin^2 A)$$

$$a^2 = c^2 - 2bc \cos A + b^2 (1)$$

$$a^2 = b^2 + c^2 - 2bc \cos A \quad \text{(Law of cosines)}$$

$$a = \sqrt{b^2 + c^2 - 2bc \cos A}$$

Example 1

A triangular piece of land has two adjacent sides of 800 feet and 500 feet. An apple orchard prevents an accurate survey of the third side. If the angle between the two adjacent sides is 86°, find the third side of the property, to the nearest foot.

Analysis

Use the Law of Cosines.

Let $m\angle A = 86°$.

Let $b = 500$.

Let $c = 800$.

Work

$$a^2 = b^2 + c^2 - 2bc \cos A$$

$$a^2 = (500)^2 + (800)^2 - 2(500)(800) \cos 86°$$

$$a^2 = 250{,}000 + 640{,}000 - 800{,}000(0.0698)$$

$$a^2 = 890{,}000 - 55{,}840$$

$$a^2 = 834{,}160$$

$$a = 913.32 \approx 913$$

Answer

913 feet

1a. In the diagram, $\overline{AC} = 24$, $\overline{CB} = 19$, and m$\angle C = 52°$. Find $\overline{AB}$. Round to the nearest tenth.

1b. The distance between Gotham City (point G) and Metropolis (point M) is 148 miles. The distance between Metropolis and Headley (point H) is 209 miles. If m$\angle GMH = 61°$, find the distance between Gotham City and Headley. Round to the nearest mile.

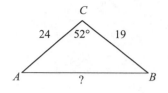

Example 2

The MacKay Construction Company is going to build a bridge over a lake fom point A to point B. A rock at point C is 350 feet from point A and 460 feet from point B. If the bridge is 534 feet long, find angle C, to the nearest degree.

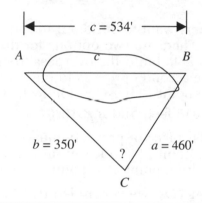

Analysis

Use the Law of Cosines to determine $\angle C$.

Let $a = 460$.

Let $b = 350$.

Let $c = 534$.

Work

$a = 460$, $b = 350$, $c = 534$:

$$a^2 = b^2 + c^2 - 2bc\cos A$$

$$(460)^2 = (350)^2 + (534)^2 - 2(350)(534)\cos A$$

$$211,600 = 122,500 + 285,156 - 373,800\cos A$$

$$211,600 = 407,656 - 373,800\cos A$$

$$-196.056 = -373,800\cos A$$

$$0.5245 = \cos A$$

Locate 0.5245 between 0.5299 and 0.5150 in the Table of Trigonometric Functions, or else use the inverse trigonometric function on the scientific calculator.

$$\begin{array}{cc} 0.5299 & 0.5245 \\ -0.5245 & -0.5150 \\ \hline 0.0054 & 0.0095 \end{array}$$

Angle	Cosine	
58°	0.5299	⎫ 0.0054
x	0.5245	⎬
59°	0.5150	⎭ 0.0095

The angle whose cosine is 0.5245 is closer to 58°.

Answer

58°

Exercises

2a.   City planners want to build a bridge over some railroad tracks. There are two options for the overpass. The first option is to build the overpass from point A to point C, and the second option is to build the overpass from point A to point B. If $\overline{AC} = 400$ feet, $\overline{BC} = 500$ feet, and $\overline{AB} = 315$ feet, find m∠ACB.

2b.   A triangular piece of property measures 148 feet from point A to point B, 210 feet from point B to point C, and 179 feet from point C to point A. Find m∠ABC.

12.4 Law of Sines

In Section 12.2, we concluded that the area of any triangle equals half the product of any two sides and the sine of the included angle. We therefore have the following possible area formulas for $\triangle ABC$, which are all equivalent:

$$\frac{1}{2}ab\sin C = \frac{1}{2}bc\sin A = \frac{1}{2}ac\sin B$$

Let's divide all these areas by $\frac{1}{2}abc$:

$$\frac{\frac{1}{2}ab\sin C}{\frac{1}{2}abc} = \frac{\frac{1}{2}bc\sin A}{\frac{1}{2}abc} = \frac{\frac{1}{2}ac\sin B}{\frac{1}{2}abc}$$

$$\frac{\sin C}{c} = \frac{\sin A}{a} = \frac{\sin B}{b}$$

The result is known as the **Law of Sines**: In any triangle, the ratio of the sine of an angle to the side opposite that angle is equal to the ratio of the sine of any other angle to the sine of the side opposite that angle.

For mnemonic reasons, we can better remember the Law of Sines if we invert all the fractions and rearrange the terms:

$$\frac{a}{\sin A} = \frac{b}{\sin B} = \frac{c}{\sin C}$$

Example 1

In triangle ABC, $m\angle C = 54°$, $m\angle A = 49°$, and $\overline{BC} = 350$ feet. Find $\overline{AB}$. Round to the nearest foot.

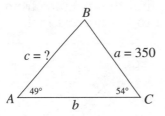

Analysis

We are given two angles and a side opposite one of these angles, so let's try the Law of Sines.

Work

$$\frac{a}{\sin A} = \frac{c}{\sin C}$$

$$\frac{350}{\sin 49°} = \frac{c}{\sin 54°}$$

$$\frac{350}{0.7547} = \frac{c}{0.8090}$$

$$0.7547c = 350 \times 0.8090$$

$$0.7547c = 283.15$$

Divide by 0.7547: $\qquad c = 375.18219 \approx 375$

Answer

375 feet

Exercises

1a. $\overline{ST}$ represents the distance from a shooter, S, to the target, T. An observer is stationed at point O. If $m\angle TOS = 63°$, $m\angle TSO = 42°$, and $\overline{TO} = 560$ feet, find the distance from the shooter to his target. Round the answer to the nearest tenth of a foot.

1b. In triangle RST, $\overline{RS} = 450$, m$\angle R = 37°$, and m$\angle T = 58°$. Find $\overline{ST}$. Round to the nearest integer.

Example 2

Shaniqua wants to determine the distance from her side of the river to a tree on the other side. She labels the spot where she is currently standing point A and the tree on the other side of the river point B. She then paces off 300 feet along the shore to point C. If m$\angle BAC = 113°$ and m$\angle ACB = 41°$, find the distance from point A to point B. Round to the nearest foot.

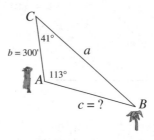

Analysis

To use the Law of Sines, we need to use two angles and the sides opposite them. We don't have all the information yet, so let's first find m$\angle B$. Then we'll be able to use the Law of Sines.

Work

The sum of the angles of a triangle add up to 180°:

$$\text{m}\angle A + \text{m}\angle B + \text{m}\angle C = 180°$$

m$\angle A = 113°$, m$\angle C = 41°$: $113° + \text{m}\angle B + 41° = 180°$

$$154° + \text{m}\angle B = 180°$$

Subtract 154: $\text{m}\angle B = 26°$

Use the Law of Sines:

$$\frac{b}{\sin B} = \frac{c}{\sin C}$$

$b = 300$, $c = ?$, m$\angle B = 26°$, m$\angle C = 41°$:

$$\frac{300}{\sin 26°} = \frac{c}{\sin 41°}$$

$$\frac{300}{0.4384} = \frac{c}{0.6561}$$

$$300(0.6561) = 0.4384c$$

$$196.83 = 0.4384c$$

$$c = 448.97 \approx 449$$

Answer

449 feet

Exercises

2a.   The distance between city A and city B is 79 miles. City C is a certain distance away from both A and B. Lines are drawn representing the distances between the cities. The measure of angle A is 33° and the measure of angle B is 82°. Find the distance between cities A and C. Round to the nearest mile.

2b.   Kelly and Delmar are standing 90 feet apart. Kelly spots a woodpecker on a tree simultaneously with Delmar. From Kelly's point of view, the woodpecker's angle of elevation is 34°. From Delmar's point of view, the woodpecker's angle of elevation is 58°. What is the distance between Delmar and the woodpecker, to the nearest tenth of a foot?

Example 3

A sea captain spots a lighthouse from his ship. The captain is 900 feet away from the foot of the lighthouse and the angle of elevation of the top of the lighthouse is 26°. Find the height of the lighthouse. Round to the nearest foot.

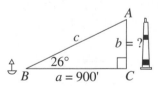

Analysis

Let $a = 900$.

Let m$\angle B = 26°$.

Find m$\angle A$ and then use the Law of Sines.

Work

Let's first find m∠A.

The sum of the angles of a △ = 180°:

$$m\angle A + m\angle B + m\angle C = 180°$$

m∠B = 26°, m∠C = 90°: $m\angle A + 26° + 90° = 180°$

$$m\angle A + 116° = 180°$$

Subtract 116: $m\angle A = 64°$

Use the Law of Sines:

$$\frac{a}{\sin A} = \frac{b}{\sin B}$$

a = 900, m∠A = 64°, m∠B = 26°:

$$\frac{900}{\sin 64} = \frac{b}{\sin 26}$$

$$\frac{900}{0.8988} = \frac{b}{0.4384}$$

$$0.8988b = 900 \times 0.4384$$

$$0.8988b = 394.56$$

Divide by 0.8988: $b = 438.98531 \approx 439$

Answer

439 feet

Exercises

3a. A wire is attached to the top of a building and makes an angle of 49° with the ground. If the wire is rooted into the ground at a point 84 feet from the foot of the building, how tall is the building? Round to the nearest foot.

3b. An airplane is at point *A* and the pilot notices two barns below, *B* and *C*. The angle of depression of barn *B* is 15°. Barn C is directly below the airplane. If the barns are nine miles apart, find the plane's altitude, to the nearest tenth of a mile.

Example 4

Route $\overline{AB}$ is a main highway. Side road $\overline{AC}$ intersects main highway $\overline{AB}$ and forms a 61° angle at the point of intersection. Side road $\overline{CB}$ also intersects main highway $\overline{AB}$. If $\overline{AC}$ = 2,600 feet and $\overline{CB}$ = 3,900 feet, what is the measure of angle B?

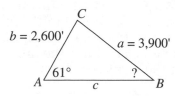

b = 2,600'
a = 3,900'
61°
?
A
c
B

Analysis

Let's try the Law of Sines.

Work

$$\frac{a}{\sin A} = \frac{b}{\sin B}$$

$a = 3{,}900$, $b = 2{,}600$, m$\angle A = 61°$:

$$\frac{3{,}900}{\sin 61°} = \frac{2{,}600}{\sin B}$$

$$\frac{3{,}900}{0.8746} = \frac{2{,}600}{\sin B}$$

$$3{,}900 \sin B = 2{,}600 \times 0.8746$$

$$3{,}900 \sin B = 2{,}273.96$$

Divide by 1,900: $\quad\quad \sin B = 0.5830666 \approx 0.5831$

Let's look at the Table of Trigonometric Functions, or else use the inverse trigonometric function on the scientific calculator.

$$\begin{array}{c} 0.5381 \quad\quad 0.5878 \\ \underline{-0.5736 \quad -0.5831} \\ 0.0095 \quad\quad 0.0047 \end{array}$$

Angle	Sine	
35°	0.5736	} 0.0095
—	0.5831	
36°	0.5878	} 0.0047

We know that 0.5831 is closer to 0.5878 than it is to 0.5736. To the nearest degree, therefore, m$\angle B = 36°$.

Answer

36°

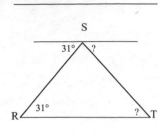

4a. In △ABC, m∠C = 43°, $\overline{AB}$ = 19, and $\overline{BC}$ = 24. Find m∠A. Round to the nearest degree.

4b. An airline pilot at point S observes two airports at points R and T. Airport T is 12 miles in a direct line from the airplane. Airport R is 7 miles in a direct line from the airplane, and the angle of depression of airport R from the airplane is 31°. Find the angle of depression of airport T from the airplane.

Test

1–8. @ 8% each = 64%

1. A 90-foot building casts a 75-foot shadow. How long is the shadow of a nearby 6-foot pole?

2. Find the distance between two trees, A and B, if the distance from a third point, C, to B is 420 feet, m∠BAC = 72°, and m∠ABC = 48°. Round to the nearest foot.

3. AB and CD intersect at E. If $\overline{AE}$ = 84, $\overline{BE}$ = 14, $\overline{EC}$ = 12, and m∠ADE = m∠ECB, find $\overline{DE}$.

4. One room has two triangular windows, ABC and DEF. $\overline{AC}$ = 16, $\overline{AB}$ = 22, $\overline{BC}$ = 18, $\overline{DE}$ = 24, $\overline{DF}$ = 33, $\overline{FE}$ = 27, m∠B = 73°, and m∠C = 29°. Find the measure of ∠D.

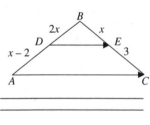

5. In the diagram, $\overline{DE}$ is parallel to $\overline{AC}$, $\overline{AD}$ = x − 2, $\overline{BD}$ = 2x, $\overline{BE}$ = x, and $\overline{CE}$ = 3. Find $\overline{AB}$.

6. A modern desk is in the shape of a triangle. If one side is 72 inches, the adjoining side is 72 inches and the angle between is 54°, find the area of the desk, to the nearest square inch.

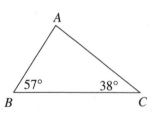

7. An airplane, at point A, has an angle of elevation of 57° when observed from point B on the ground and an angle of elevation of 38° when observed from point C on the ground. Point B is 1,900 feet away from point C. Find $\overline{AB}$, to the nearest foot.

8. The area of triangular rudder ABC on a boat is 375 square inches. If $\overline{AB}$ = 42 inches and $\overline{BC}$ = 48 inches, find the measure of angle B.

9–12. @ 9% each = 36%

9. The angle of elevation of the top of a building from a point on the ground is 26°. If the distance from that point to the base of the building is 740 feet, find the height of the building. Round to the nearest foot.

10. A direct road joining town A to town B runs 57 miles. Another direct road from town B to town C runs

81 miles while a third direct route from town C to town A is 63 miles. Find the measure of angle ACB.

11.   On a particular day, if an imaginary line is drawn from the center of the earth to the center of the sun, the distance is 91 million miles. At the same time, if an imaginary line is drawn from the center of Jupiter to the center of the sun, the distance is 450 million miles. If the included angle at the sun was 89°, find the distance between the earth and Jupiter, to the nearest million miles.

12.   A direct road from city A to city B runs 24 miles. Another direct road from City B to city C runs 17 miles while a third direct road from city C to city A runs 15 miles. What is the angle formed by the intersection of roads $\overline{AB}$ and $\overline{BC}$?

13
Differentiation

The ancient Greek philosopher Zeno of Elea (c. 460 B.C.E.), in his paradox, touched upon the fundamental idea of calculus, the idea of infinitesimal increments:

Achilles and a tortoise race along a straight track. Achilles is much faster, but the tortoise is given a short lead. The tortoise reaches point A. In order to pass the tortoise, Achilles reaches point A, but by that time the tortoise

has reached another point, point *B*. Once Achilles reaches point *B*, the tortoise is on to point *C*, and so forth. Achilles can never surpass the tortoise.

Although we cannot solve this paradox within the given parameters, Zeno's fundamental contribution to calculus was to divide an interval into smaller and smaller partitions.

Zeno's suggestion remained an interesting puzzle with no practical applications until some 1,200 years later when, in 1635, Bonaventura Cavalieri, a student of Galileo, proposed a theory of calculus fairly close to our own concept. In Cavalieri's scheme, a point generates a line and line segments are added together to form an area. A moving line generates a plane while plane segments are added together to form a volume.

Although Zeno, Cavalieri, and even Galileo approached modern calculus, it was the great English mathematician Isaac Newton (1642–1727) and the German mathematician Gottfried Wilhelm Leibniz (1646–1716) who independently packaged contemporary calculus. They used the earlier formulas of algebra and trigonometry and built upon them.

Both Newton and Leibniz went beyond the constants used in arithmetic and the variables used in algebra. They wondered about describing falling objects where the speed was constantly increasing or leaking containers when the volume was constantly decreasing. In the years between 1664 and 1666, Newton unified various earlier mathematical techniques and used differentiation to find maximum and minimum points, areas, lengths of curves, and tangents. With a leap of mathematical insight, he realized that the integration of a function, or the determination of the area under a curve, was simply the reverse of differentiation.

By 1676, Leibniz was suggesting the same sort of mathematical ideas that Newton had developed 10 years earlier. Leibniz could find the sums and differences of rational, irrational, algebraic, or transcendental functions. He developed the notations *dy* and *dx* for the smallest possible differences in *x* and *y* (differentials).

Whereas Newton preferred to illustrate his calculus using velocity as an example, Leibniz preferred to work with infinitesimal amounts in his illustrations. Today we use the symbols for derivatives and integrals suggested by Leibniz, but we employ Newton's formulas and theorems.

In the eighteenth century, the French/Italian mathematician and astronomer Joseph Louis Lagrange (1736–1813) resurrected Zeno's illustration and began to use an infinite series in his approach to calculus.

Although Newton, Leibniz, and Lagrange introduced the modern tools of calculus, all of them were unable to present a rigorous explanation of their mathematics. It was left to the French mathematician Augustin Louis Cauchy (1789–1857) to provide a rigorous proof of calculus. In 1821, he proposed a proof based on finite amounts, the idea of limit, and the convergence of an infinite series.

13.1 Limits of Functions

The concept of the limit of a function is fundamental to the study of calculus.

We're all familiar with formulas. A formula links together variables and constants in some sort of relationship. Given certain information, we can go on to find the value of an unknown quantity in the relationship.

For example, let's take the general formula for a linear equation:

$$y = 2x + 4$$

where 2 and 4 are constants and x and y are variables. In this equation, once we are given the value of x, we can then determine the value of y. In other words, the value of y depends upon the value of x. The *independent variable* is x, while y is the *dependent variable*.

If we have two related variables, x and y, such that whenever we're given a specific value for x, there is one and only one corresponding value of y, we say that y is a function of x:

$$y = f(x)$$

The function $y = x^2 + 3x - 1$ tells us to perform a series of operations on the x variable in order to determine the y variable. The same function may be written in the form $f(x) = x^2 + 3x - 1$.

Calculus superimposed the concept of limit onto the function.

$$\lim_{x \to a} f(x) = A$$

Here, $f(x)$ gets closer and closer to A as x gets closer and closer to a.

Theorems on Limits

In the following theorems, $\lim_{x \to a} f(x) = A$ and $\lim_{x \to a} g(x) = B$.

1. If $f(x) = A$ (a constant), then $\lim_{x \to a} f(x) = A$.

2. $\displaystyle\lim_{x\to a} c\cdot f(x) = c\,\lim_{x\to a} f(x) = c\cdot A.$

3. $\displaystyle\lim_{x\to a}\{f(x)\pm g(x)\} = \lim_{x\to a} f(x)\pm\lim_{x\to a} g(x) = A\pm B.$

4. $\displaystyle\lim_{x\to a}\left[\frac{f(x)}{g(x)}\right] = \frac{\displaystyle\lim_{x\to a} f(x)}{\displaystyle\lim_{x\to a} g(x)} = \frac{A}{B},\qquad B\neq 0.$

Example 1

Find $\displaystyle\lim_{x\to 3} 77.$

Analysis

The variable x is irrelevant to the constant, 77.

Answer

$\displaystyle\lim_{x\to 3} 77 = 77.$

1a. Find $\displaystyle\lim_{x\to 2} 98.$ 1b. Find $\displaystyle\lim_{x\to 5} 7.2.$

Exercises

Example 2

Find $\displaystyle\lim_{x\to 4} 6x.$

Work

$$\lim_{x\to 4} 6x = 6\lim_{x\to 4} x = 6\cdot 4 = 24$$

Answer

24

2a. Find $\displaystyle\lim_{y\to 6} 7y.$ 2b. Find $\displaystyle\lim_{x\to 4} 12x.$

Exercises

Example 3

Find $\displaystyle\lim_{x\to 3}(5x-8).$

Work

$$\lim_{x\to 3}(5x-8) = \lim_{x\to 3} 5x - \lim_{x\to 3} 8 = 5\lim_{x\to 3} x - 8 = 5\cdot 3 - 8 = 7$$

Answer

7

3a. Find $\lim\limits_{x\to 1}(6x+7)$.   3b. Find $\lim\limits_{y\to 2}(7y-5)$.

Example 4

Find $\lim\limits_{x\to 4}\dfrac{x^2-16}{x-4}$.

Work

$$\lim_{x\to 4}\frac{x^2-16}{x-4}=\frac{4^2-16}{4-4}=\frac{0}{0}$$

Division by zero is undefined, so we have not yet found the limit for this function.

Let's try factoring.

$$\lim_{x\to 4}\frac{x^2-16}{x-4}=\lim_{x\to 4}\frac{(x-4)(x+4)}{(x-4)}=\lim_{x\to 4}x+4=8$$

Answer

8

4a. Find $\lim\limits_{x\to 3}\dfrac{x^2-9}{x-3}$.   4b. Find $\lim\limits_{x\to 1}\dfrac{x-1}{x^2-1}$.

Example 5

Find $\lim\limits_{x\to 0}\dfrac{x}{\sqrt{x+2}-\sqrt{2}}$.

Analysis

If we substitute 0 for x directly, we get zero in the denominator, which results in an undefined answer, so we'll have to try some other method. Let's multiply both denominator and numerator by $\sqrt{x+2}+\sqrt{2}$.

Work

$$\lim_{x\to 0}\frac{x}{\sqrt{x+2}-\sqrt{2}}\cdot\frac{\sqrt{x+2}+\sqrt{2}}{\sqrt{x+2}+\sqrt{2}}=\lim_{x\to 0}\frac{x\left(\sqrt{x+2}+\sqrt{2}\right)}{(x+2)-2}$$

$$=\lim_{x\to 0}\frac{x\left(\sqrt{x+2}+\sqrt{2}\right)}{x}$$

$$=\lim_{x\to 0}\left(\sqrt{x+2}+\sqrt{2}\right)$$

$$=\lim_{x\to 0}\left(\sqrt{2}+\sqrt{2}\right)=2\sqrt{2}$$

Answer

$2\sqrt{2}$

5a. Find $\lim\limits_{x\to 0}\dfrac{x}{\sqrt{x+3}-\sqrt{3}}$.   5b. Find $\lim\limits_{x\to 0}\dfrac{x}{\sqrt{x+7}-\sqrt{7}}$.

The best way to study derivatives is via a geometric interpretation. Curved line l_1 is a geometric picture of the function $y = f(x)$. x and $(x + h)$ are two points on the x-axis, and $f(x)$ and $f(x + h)$ are the corresponding y coordinates. h is a small positive number so that $(x + h)$ is just a small distance to the right of x. Line l_2 connects the two points $(x, f(x))$ and $(x + h, f(x + h))$. The slope of line l_2 is

$$\frac{f(x+h)-f(x)}{(x+h)-x} = \frac{f(x+h)-f(x)}{h} = \frac{\Delta y}{\Delta x}$$

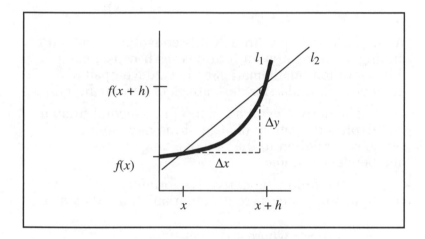

Since we're dividing $[f(x + h) - f(x)]$ by h, the slope of line l_2 represents the average rate of change of f with respect to x.

If we make h smaller and smaller and closer to 0 and we keep drawing lines connecting points $(x, f(x))$ and $(x + h, f(x + h))$, the lines get closer and closer to tangent line t.

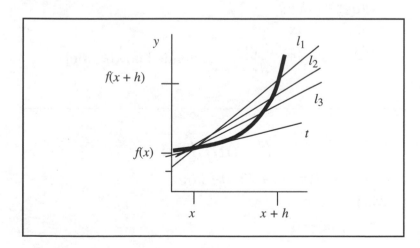

The slope of the tangent is

$$\frac{f(x+h)-f(x)}{h}$$

If we let h approach zero, we obtain the definition of the derivative of a function f at x.

$$f'(x) = \lim_{h \to 0} \frac{f(x+h)-f(x)}{h}$$

The derivative of f at x is, therefore, also the slope of the tangent line to the curve at $(x, f(x))$.

If $f(x)$ is our original function, its derivative may be represented by any of the following notations:

$$f'(x), \qquad \frac{dy}{dx}, \qquad y', \qquad \frac{d}{dx}(f(x))$$

We don't have to go through this convoluted method of finding the derivative in future cases because, using this definition, mathematicians have developed a number of formulas for derivatives of various functions.

The derivative of a function is a second function that depicts the rate of change of the dependent variable in relation to the rate of change of the independent variable.

In the following derivative formulas, a is a constant, and y and z are differentiable functions of x.

$$\frac{d(a)}{dx} = 0, \qquad \text{where } a \text{ is a constant}$$

$$\frac{d(yz)}{dx} = y\frac{d(z)}{dx} + z\frac{d(y)}{dx} \qquad \text{[product rule]}$$

$$\frac{d(x)}{dx} = 1$$

$$\frac{d(x^m)}{dx} = mx^{m-1} \qquad \text{[power rule]}$$

$$\frac{d(ay)}{dx} = a\frac{d(y)}{dx}$$

$$\frac{d(y^m)}{dx} = my^{m-1}\frac{d(y)}{dx} \qquad \text{[extended power rule]}$$

Example 1

Let $f(x) = 19$. Find $f'(x)$.

Work

$$f'(x) = 0.$$

Answer

0

1a. Find the derivative of $f(x) = 38$.

1b. Let $f(x) = 564$. Find $f'(x)$.

Exercises

Example 2

Find the derivative of $f(x) = 7x$.

Work

$$\frac{d(7x)}{dx} = 7$$

Answer

7

2a. If $f(x) = 45x$, find $f'(x)$.

2b. Find the derivative of $f(x) = 23x$.

Exercises

Example 3

Find the derivative of $f(x) = 5x^3$.

Work

$$f'(x) = 3 \cdot 5x^2 = 15x^2$$

Answer

$15x^2$

3a. Let $f(x) = 6x^4$. Find $f'(x)$.

3b. If $f(x) = 4x^5$, find $f'(x)$.

Exercises

Example 4

Differentiate $y = 6x^4 - 5x^3 + x^2 - 6x + 7$.

Work

$$\frac{dy}{dx} = 4 \cdot 6x^3 - 3 \cdot 5x^2 + 2x - 6 + 0 = 24x^3 - 15x^2 + 2x - 6$$

Answer

$$24x^3 - 15x^2 + 2x - 6$$

Exercises

4a. Find the derivative of $f(x) = 3x^4 - 2x^3 + x^2 - 6$.

4b. Differentiate $y = -2x^5 + 8x^4 - 3x^3 + 7x^2 - 5x + 9$.

Example 5

Find the derivative of $f(x) = (6x^2 + 2x)(3x - 1)$.

Analysis

First, multiply the terms and then differentiate.

Work

$$f(x) = (6x^2 + 2x)(3x - 1)$$
$$= 18x^3 - 6x^2 + 6x^2 - 2x = 18x^3 - 2x$$
$$f'(x) = 3 \cdot 18x^2 - 2 = 54x^2 - 2$$

Answer

$$54x^2 - 2$$

Exercises

5a. Find the derivative of $y = (2x + 3)(4x^3 - 2x^2)$.

5b. Differentiate $f(x) = (4x^2 + 7)(5x^4 - 2x + 9)$.

Example 6

Differentiate $y = (2x^2 - 4x)^2$.

Work

Let $u = 2x^2 - 4x$.

By the power rule, $\dfrac{du}{dx} = 4x - 4$

$$u^2 = (2x^2 - 4x)^2$$

Let $y = u^2$.

By the extended power rule,

$$\frac{dy}{dx} = 2u\frac{du}{dx}$$

$$= 2(2x^2 - 4x)(4x - 4)$$

$$= (4x^2 - 8x)(4x - 4)$$

$$= 16x^3 - 16x^2 - 32x^2 + 32x$$

$$= 16x^3 - 48x^2 + 32x$$

Answer

$$y' = 16x^3 - 48x^2 + 32x$$

6a. Differentiate $y = (4x^3 + 5x^2 - 3)^2$.

6b. Find the derivative of $f(x) = (2x^4 - 3x^3 + 2)^2$.

Exercises

Example 7

Find the derivative of $y = \sqrt{5 - 3x}$.

Work

$$y = (5 - 3x)^{\frac{1}{2}}$$

Let $u = 5 - 3x$.

By the power rule,

$$\frac{du}{dx} = -3$$

$$u^{\frac{1}{2}} = (5 - 3x)^{\frac{1}{2}}$$

Let $y = u^{\frac{1}{2}}$.

By the extended power rule,

$$\frac{dy}{dx} = \frac{1}{2}u^{-\frac{1}{2}}\frac{du}{dx}$$

$$= \frac{1}{2}(5 - 3x)^{-\frac{1}{2}}(-3)$$

$$= \frac{-3}{2\sqrt{5 - 3x}}$$

Answer

$$\frac{-3}{2\sqrt{5 - 3x}}$$

7a. If $f(x) = \sqrt{4x + 7}$, find $f'(x)$.

7b. Differentiate $y = \sqrt{5x - 4}$.

Exercises

Example 8

A rock is dropped into a lake. The ripples spread outward in a circle. If the radius of the circle increases at the rate of 2.7 ft/sec, how fast is the area of the circle increasing when the radius is 9 feet? Round to the nearest tenth of a foot.

Analysis

The radius and the area are linked by the formula $A = \pi r^2$, where A = the area of the circle, $\pi = 3.14$, and r = the radius of the circle.

Work

The radius of the circle is increasing, dr, with respect to the increase in time, dt, at the rate of 2.7 ft/sec:

$$\frac{dr}{dt} = 2.7$$

The area of the circle is increasing, dA, with respect to the increase in time, dt:

$$\frac{dA}{dt} = \frac{d(\pi r^2)}{dt} = 2\pi r \frac{dr}{dt}$$

$\pi = 3.14$, $r = 9$, $\dfrac{dr}{dt} = 2.7$:
$$\frac{dA}{dt} = 2(3.14)(9)(2.7)$$
$$= 152.604 \approx 152.6$$

Answer

The area of the circle is increasing at the rate of 152.6 ft²/sec.

Exercises

8a.   A side of a cube is increasing at the rate of 0.1 inch per second. Find the rate of increase in the volume when a side is 2 inches. $V = s^3$.

8b.   Jessie is pumping up a balloon and the radius is increasing at the rate of 0.5 inch per second. Find the rate of change of the volume when the radius is 2 inches. Round to the nearest tenth of a cubic inch. $V = (4/3)\,\pi r^3$ and $\pi = 3.14$.

Example 9

The cost of producing n items is given by the function $C(n) = 0.003n^2 + 18n - 400$ dollars, where n = the number of items. Find the rate of change in the cost when 300 items are produced.

Analysis

Find the derivative of $C(n)$ with respect to n and then let $n = 300$.

Work

The original function: $C(n)$ $\quad = 0.003n^2 + 18n - 400$

The derivative: $\quad C'(n) \quad = 0.006n + 18$

$n = 300$: $\quad\quad C'(300) = 0.006(300) + 18$
$\quad\quad\quad\quad\quad\quad\quad\quad = 19.8$

Answer

$19.80 per item

9a. A hotel determines that if they have n number of guests, their profit is $P(n) = 200 - 50n + (2/5)n^2$. Find the change in the rate of profit when the hotel books 150 guests.

9b. The income of a company is related to the number of salespeople they employ, s, by the formula $I(s) = \$10{,}000 - s^2 + 20{,}000s$. Find the change in the rate of income when the company employs five salespeople.

Exercises

Example 10

Demographers have developed a formula to represent the future population of Middletown, USA: $P(y) = y^3 + 6y^2 + 124{,}000$, where y represents years from now. Find the rate of change per year 34 years from now.

Analysis

First, find the rate of change in the population with respect to years, $P'(y)$. Then substitute 34 years for y in the answer.

Work

Original formula: $P(y) = y^3 + 6y^2 + 124{,}000$

Differentiate with respect to

years, y: $P'(y) = 3y^2 + 12y$

$y = 34$: $P'(34) = 3(34)^2 + 12(34)$
    $= 3{,}468 + 408$

    $P'(34) = 3{,}876$

Answer

In 34 years the population will increase at the rate of 3,876 people per year.

Exercises

10a.   The monthly pounds of garbage generated by a community is related to the number of persons in that community, n, by the formula $G(n) = 1.5n + 3\sqrt{2n}$. Find the rate of change in the amount of garbage generated with 50 people in the community.

10b.   The milligram growth of bacteria per hour, t, is given by the function $B(t) = 1{,}000 + 4t^2 - 5t$. Find the rate of growth of bacteria when $t = 6$.

13.3 Slope of a Curve at a Specific Point

Let's graph the equation $y = 2x^2 + 2x - 3$ and then determine the slope of the graph at $x = 1$.

x	$2x^2 + 2x - 3$	y
−3	$2(-3)^2 + 2(-3) - 3$	9
−2	$2(-2)^2 + 2(-2) - 3$	1
−1	$2(-1)^2 + 2(-1) - 3$	−3
0	$2(0)^2 + 2(0) - 3$	−3
1	$2(1)^2 + 2(1) - 3$	1
2	$2(2)^2 + 2(2) - 3$	9

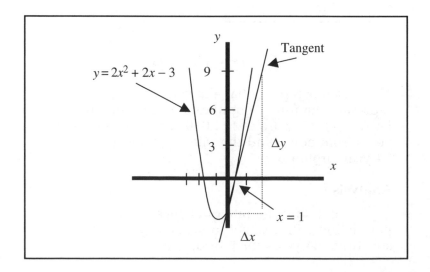

The slope of the line tangent to the graph $y = 2x^2 + 2x - 3$ at $x = 1$ is $\Delta y / \Delta x$.

However, since $dy/dx = \lim\limits_{\Delta x \to 0}(\Delta y / \Delta x)$ (provided this limit exists), we must differentiate to obtain the tangent to the curve. After we obtain the equation of this tangent line, we just substitute the value for x at that point to find the slope of the line.

Example 1

Find the slope of the tangent to the graph of the function $y = 2x^2 + 2x - 3$ at the point $x = 1$.

Analysis

Differentiate and then substitute 1 for x.

Work

$$y = 2x^2 + 2x - 3$$

Find dy/dx:
$$\frac{dy}{dx} = 4x + 2$$

Find the slope at $x = 1$:
$$\frac{dy}{dx} = 4(1) + 2 = 6$$

Answer

6

Exercises

1a. Find the slope of the tangent to the curve of the graph of the function $y = 3x^2 - x + 5$ at the point $x = 3$.

1b. Find the slope of the line tangent to the graph of $f(x) = -x^3 + 4x - 8$ at the point $x = -1$.

Example 2

Find the slope of the line tangent to the graph of $y = x^2 + 2x$ at the point where $y = 3$.

Analysis

Substitute 3 for y in the original function. Find x. Then find the derivative of the original function and substitute the value of x into the derivative.

Work

Function:	$y = x^2 + 2x$
Let $y = 3$:	$3 = x^2 + 2x$
Subtract 3:	$0 = x^2 + 2x - 3$
Symmetric Property:	$x^2 + 2x - 3 = 0$
Factor:	$(x + 3)(x - 1) = 0$

$$x + 3 = 0 \qquad x - 1 = 0$$
$$x = -3 \qquad x = 1$$

There are two tangents to the graph, at $x = 3$ and $x = -1$, so find the derivative of the original function and substitute for x.

Function:	$y = x^2 + 2x$
Derivative:	$y' = 2x + 2$
$x = -3$:	$y' = 2(-3) + 2 = -6 + 2 = -4$
$x = 1$:	$y' = 2(1) + 2 = 2 + 2 = 4$

Answer

$$\text{slope} = -4 \text{ at } x = -3, \text{ slope} = 4 \text{ at } x = 1$$

Exercises

2a.   Find the slope to the curve of the graph of the function $y = x^2 + 6x + 12$ at the point $y = 4$.

2b.   Find the slope of the line tangent to the graph of $y = x^2 + 1x + 5$ at the point $y = 25$.

13.4 Second Derivatives

If $y = f(x)$ is differentiable, its derivative, y' or $f'(x)$, is called the first derivative. If the first derivative is differentiable, its derivative, y'' or $f''(x)$, is called the second derivative.

To obtain the first derivative, let's first use the extended power rule:

$$\frac{d(y^m)}{dx} = my^{m-1}\frac{d(y)}{dx}$$

Then, to obtain the second derivative, we'll just use the extended power rule a second time.

Example 1

Find the second derivative of $y = 7x$.

Work

Original function: $y = 7x$

First derivative: $\qquad y' = 7$

Second derivative: $\qquad y'' = 0$

Answer

$\qquad 0$

1a. Find the second derivative of $y = 9x$.

1b. If $f(x) = 82x$, find $f''(x)$.

Example 2

Find the second derivative of $f(x) = 8x^3$.

Work

Original function: $\qquad f(x) = 8x^3$

First derivative: $\qquad f'(x) = 3 \cdot 8x^2 = 24x^2$

Second derivative: $\qquad f''(x) = 2 \cdot 24x^1 = 48x$

Answer

$\qquad 48x$

2a. Find the second derivative of $f(x) = 18x^2$.

2b. $y = 7x^4$. Find y''.

Example 3

Find the second derivative of $y = 7x^3 + 3x^2 - 5x + 12$.

Work

Original function: $\qquad y = 7x^3 + 3x^2 - 5x + 12$

First derivative: $\qquad y' = 3 \cdot 7x^2 + 2 \cdot 3x - 5$
$\qquad\qquad\qquad\quad = 21x^2 + 6x - 5$

Second derivative: $\qquad y'' = 2 \cdot 21x + 6 = 42x + 6$

Answer

$\qquad 42x + 6$

3a. Find the second derivative of $f(x) = 12x^3 - 2x^2 + 4x - 9$.

3b. If $f(x) = 7x^4 - 2x^3 + 7x^2 - 8x - 11$, find $f''(x)$.

Example 4

Find the second derivative of $f(x) = (5/6)x^{2/3}$.

Work

Original function: $f(x) = (5/6)x^{2/3}$

First derivative: $f'(x) = (2/3)(5/6)x^{-1/3}$
$= (5/9)x^{-1/3}$

Second derivative: $f''(x) = (-1/3)(5/9)x^{-4/3}$
$= (-5/27)x^{-4/3}$

Answer

$(-5/27)x^{-4/3}$

4a. Find the second derivative of $f(x) = (2/3)x^{2/5}$.

4b. If $f(x) = (3/4)x^{-1/3}$, find $f'(x)$.

Example 5

Find the second derivative of $y = (3x^3 - 2x)^2$.

Work

Original function: $y = (3x^3 - 2x)^2$

First derivative: $\dfrac{dy}{dx} = (2)(3x^3 - 2)(9x^2)$

$= 18x^2(3x^3 - 2)$

$= 54x^5 - 36x^2$

Second derivative: $\dfrac{d^2y}{dx^2} = 270x^4 - 72x$

Answer

$270x^4 - 72x$

5a. Find the second derivative of $y = (5x^2 + 7)^2$.

5b. Find the second derivative of $(7x^3 + 2)^2$.

Example 6

The number of ants in a colony is represented by $A(t) = 5t^3 + 900t^2 - 400t$, where t = days from now. $A'(t)$ depicts the rate of growth of the colony. $A''(t)$ represents the change in the rate of growth of the colony. Find the change in the rate of growth in the colony in 40 days, $A''(40)$.

Analysis

Differentiate twice and substitute 40 for t in the second differential.

Work

Original function:	$A(t) = 5t^3 + 90t^2 - 400t$
First derivative:	$A'(t) = 15t^2 + 180t - 400$
Second derivative:	$A''(t) = 30t + 180$
$t = 40$:	$A''(40) = 30(40) + 180$
	$= 1,200 + 180 = 1,380$

Answer

1,380 ants per day

Exercises

6a. $P(t) = t^3 + 5t^2 + 30$ represents the number of pounds of pollutants dumped into a local river in t days. $P'(t)$ is the rate of growth of pollutants while $P''(t)$ represents the change in the rate of growth of the pollutants. Find $P''(4)$.

6b. The number of new words learned in a foreign language is indicated by the formula $W(t) = 2t^3 - 4t^2 + 6t$, where t indicates days. $W'(t)$ represents the rate of growth of learning, while $W''(t)$ represents the change in the rate of growth of learning. Find the change in the rate of growth of learning in three days.

13.5 Maximum and Minimum Points

Whenever $\Delta y/\Delta x > 0$, the slope is positive and y increases as x increases.

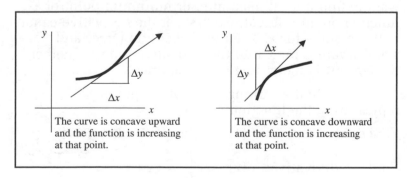

The curve is concave upward and the function is increasing at that point.

The curve is concave downward and the function is increasing at that point.

Whenever $\Delta y/\Delta x < 0$, the slope is negative and y decreases as x increases.

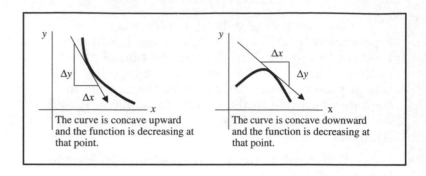

The curve is concave upward and the function is decreasing at that point.

The curve is concave downward and the function is decreasing at that point.

In the graph directly below at the left, the slope of tangent t_1 is negative and the slope of tangent t_2 is positive. This indicates that there is a point in between where the slope changes from $-$ to $+$. That point is (a, b) where the slope of tangent t_3 to that point is 0.

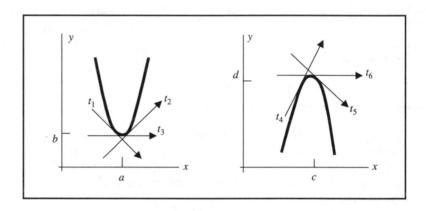

In the graph directly above at the right, the slope of tangent t_4 is positive, and the slope of tangent t_5 is negative. This pattern indicates that there is a point in between where the slope changes from $+$ to $-$. That point is (c, d) where the slope of tangent t_6 to that point is 0.

First-Derivative Test

Since the tangents to both minimum and maximum points have slopes of zero, whenever we want to find a local maximum or minimum point of a function on an interval, we first set the derivative equal to 0 and solve for x. Then we substitute larger and smaller values for x into the derivative to test whether the tangent is increasing or decreasing.

Let's find a turning point of the graph representing the function $y = 2x^2 + 4x - 3$.

Function: (1) $y = 2x^2 + 4x - 3$

Find the derivative: (2) $\dfrac{dy}{dx} = 4x + 4$

Set the derivative equal to

0: $\qquad 4x + 4 = 0$

Factor: $\qquad 4(x + 1) = 0$

Divide by 4: $\qquad x + 1 = 0$

Subtract 1: $\qquad x = -1$

To find y, go back to (1) and substitute -1 for x:

$\qquad\qquad$ (1) $\qquad y = 2x^2 + 4x - 3$

$x = 1$: $\qquad\qquad\qquad y = 2(-1)^2 + 4(-1) - 3$

$\qquad\qquad\qquad\qquad y = -5$

The turning point is $(-1, -5)$.

$\qquad$ Is that turning point a local maximum or a minimum point? Once we have that point, we can then substitute values for x before and after that point to determine the slopes of the tangents.

$\qquad$ Let's go back to (2), the derivative, and substitute values for x a bit smaller and greater than -1.

$\qquad\qquad$ (2) $\qquad \dfrac{dy}{dx} = 4x + 4$

$x = -1.5$: $\qquad \dfrac{dy}{dx} = 4(-1.5) + 4 = -2$ (The slope of

$\qquad\qquad\qquad$ the tangent is negative, and

$\qquad\qquad\qquad$ the function is decreasing.)

$x = -0.5$: $\qquad \dfrac{dy}{dx} = 4x + 4$

$\qquad\qquad \dfrac{dy}{dx} = 4(-0.5) + 4 = +2$ (The slope of

$\qquad\qquad\qquad$ the tanget is positive, and

$\qquad\qquad\qquad$ the function is increasing.)

$\qquad$ Let's sketch the tangents near the point $(-1, -5)$ and then determine whether that point is a minimum or a maximum.

$\qquad$ We can clearly see that $(-1, -5)$ is a minimum point.

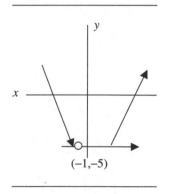

$\qquad$ On the other hand, if we obtain the following sketch of the tangents, we know that we have a maximum point.

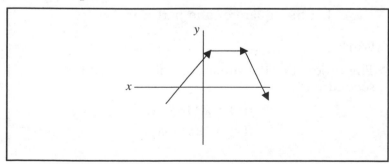

Second-Derivative Test

In our analysis of the first derivative, we have discovered that when a function increases, its first derivative is positive, and when it decreases, the first derivative is negative.

When the second derivative of a function is positive over an interval, the first derivative is increasing and the graph of the function is concave upward—a local minimum.

Analogously, when the second derivative of a function is negative over an interval, the first derivative is decreasing and the graph of the function is concave downward—a local maximum.

Let's return to our function $y = 2x^2 + 4x - 3$. Recall that the critical point is $(-1, -5)$, the graph of the function is concave upward, and the point is a local minimum.

To determine whether this is a local minimum or a maximum point, use the second derivative test.

Function: (1) $f(x) = 2x^2 + 4x - 3$

First derivative: (2) $f'(x) = 4x + 4$

Second derivative: (3) $f''(x) = 4$

Since $4 > 0$, the graph is concave upward—a minimum point.

Summary of Local Minimum/ Maximum Point

The local minimum or local maximum occurs when the slope is zero, which we find by setting the first derivative to 0. When the second derivative is negative, the function is a local maximum. When the second derivative is positive, the function is at a local minimum.

Example 1

Separate the number 18 into two parts such that the product of the square of one part and the second part is at a maximum.

Analysis

Let x = one part.

Let $18 - x$ = the second part.

Let the square of one part = x^2.

Work

The product of the square of one part and the second part:

$$f(x) = x^2(18 - x)$$
$$f(x) = 18x^2 - x^3$$

To find the local minimum or maximum, first find the derivative of the function:

$$f'(x) = 36x - 3x^2$$

Set the derivative equal to zero:

$$36x - 3x^2 = 0$$

Factor: $\qquad\qquad\qquad\qquad 3x(12 - x) = 0$

$$3x = 0 \qquad\qquad 12 - x = 0$$
$$x = 0 \qquad\qquad -x = -12$$

Reject $x = 0$ because $x^2(18 - x) =$ $\qquad\qquad x = 12$
    $0(18) = 0$, not a maximum.

To determine whether $x = 12$ is a minimum or a maximum point, let's find the second derivative.

$$f''(x) = 36 - 6x$$

$x = 12$: $\qquad f''(12) = 36 - 6(12) = 36 - 72 = -36$

When the second derivative is negative, the function is a local maximum. The first part is $x = 12$, the second part is $18 - x = 18 - 12 = 6$.

Answer

    12, 6

1a.   Separate 15 into two parts such that the product of the two parts is a local maximum.

1b.   Separate 20 into two parts such that the product of the cube of one part and the second part is a local maximum.

Exercises

Example 2

The local school board has 2,000 feet of fencing, and the members want to construct a playground area for students. What are the dimensions for enclosing the largest possible rectangular area?

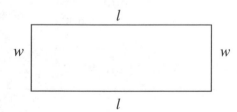

Analysis

Let A = area.

Let l = length.

Let w = width.

Let P = perimeter.

First, find l in terms of w. Then find dA/dw, set the result equal to zero, and find w. Last of all, find d^2A/dw^2 to determine whether the point is a local minimum or maximum.

Work

$$P = 2l + 2w$$

$P = 2,000$: $\quad\quad\quad\quad\quad\quad\quad 2,000 = 2l + 2w$

Symmetric Property: $\quad\quad\quad 2l + 2w = 2,000$

$$2l = 2,000 - 2w$$

$$l = 1,000 - w$$

A = area, l = length, w = width: $\quad A = lw$

$l = 1,000 - w$: $\quad\quad\quad\quad\quad\quad A = (1,000 - w)w$

$$A = 1,000w - w^2$$

Derivative: $\quad\quad\quad\quad\quad\quad\quad \dfrac{dA}{dw} = 1,000 - 2w$

Set the derivative

equal to 0: $\quad\quad\quad\quad\quad\quad 1,000 - 2w = 0$

$$-2w = -1,000$$

$$w = 500$$

To determine whether this is a minimum or a maximum point, let's find the second derivative.

$$\frac{d^2A}{dw^2} = -2$$

When the second derivative is negative, the function is a local maximum.

Now, let's find l. Go back to the formula for the perimeter.

$$P = 2l + 2w$$

$P = 2,000$, $w = 500$: $\quad 2,000 = 2l + 2(500)$

$$2,000 = 2l + 1,000$$

$$1,000 = 2l$$

$$500 = l$$

$$l = 500$$

Answer

Length = 500 ft, Width = 500 ft

Exercises

2a. Lucy wants to enclose a rectangular area with 500 feet of fencing. What are the dimensions of the largest possible area?

2b. Henry has 900 feet of fencing. He wants to enclose a rectangular piece of property, but he only has to use fencing for three sides because he will use a wall for the fourth side. What are the dimensions of the largest area he can enclose?

Example 3

We have a 20″ × 20″ rectangular piece of cardboard, which we wish to covert into an open box. So we cut out four squares, one from each corner and fold up the edges. How many inches should each square be to maximize the volume of the box?

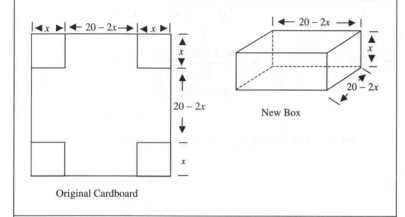

Original Cardboard   New Box

Analysis

Let x = each side of the square.

Let V = volume.

Let l = length.

Let w = width.

Let h = height.

Volume = Length × Width × Height

length = $20 - 2x$

width = $20 - 2x$

height = x

Work

$$V = lwh$$
$$V = (20 - 2x)(20 - 2x)(x)$$
$$V = (400 - 80x + 4x^2)x$$
$$V = 400x - 80x^2 + 4x^3$$

Find the first derivative: $V' = 400 - 160x + 12x^2$

Set the first derivative equal to zero:

$$12x^2 - 160x + 400 = 0$$

$$4(3x^2 - 40x + 100) = 0$$

$$3x^2 - 40x + 100 = 0$$

Use the quadratic equation:

$$x = \frac{-b \pm \sqrt{b^2 - 4ac}}{2a}$$

$a = 3, \; b = -40,$
$c = 100$:

$$x = \frac{-(-40) \pm \sqrt{(-40)^2 - 4(3)(100)}}{2(3)}$$

$$= \frac{40 \pm \sqrt{1,600 - 1,200}}{6}$$

$$= \frac{40 \pm \sqrt{400}}{6} = \frac{40 \pm 20}{6}$$

$$x_1 = \frac{40 + 20}{6} = \frac{60}{6} = 10$$

$$x_2 = \frac{40 - 20}{6} = \frac{20}{6} = 3\frac{2}{3}$$

To determine a minimum or maximum, find the second derivative:

$$V' = 400 - 160x + 12x^2$$

$$V'' = -160 + 24x$$

Substitute 10 for x: $\quad V'' = -160 + 24(10)$
$$= -160 + 240 = 80$$

When the second derivative is positive, the function is at a minimum.

Substitue $3\frac{2}{3}$
for x: $\qquad V'' = -160 + 24x = -160 + 24\left(3\frac{2}{3}\right)$

$$= -160 + 88 = -72$$

When the second derivative is negative, the function is at a maximum.

Answer

      Each side of the cut square should be $3\frac{2}{3}$ inches.

Exercises

3a.   We plan to construct an open box out of a 16-inch by 16-inch piece of cardboard by removing a square from each corner. What are the dimensions of the square that should be cut from each corner to make a box with the maximum volume?

3b.   Janice is going to construct a box with an open top by cutting a square from each corner of a piece of cardboard, 14 inches by 14 inches. Find a side of the cut square that will form a box with the greatest capacity.

Example 4

The Ajax Corporation produces computers. It costs $(600 + n^2 + 4n)$ million dollars to manufacture n million computers. If the corporation sells n million computers at $\$(200 - 3n)$ each, how many should Ajax produce to maximize profits?

Analysis

Let n = number of computers.

Let P = profit.

Let TC = total cost = $600 + n^2 + 4n$, where n is in millions of dollars.

Let TSP = total selling price = $n(200 - 3n)$, in millions of dollars.

Work

$$P = TSP - TC$$

$$P(n) = n(200 - 3n)$$
$$- (600 + n^2 + 4n)$$

$$P(n) = 200n - 3n^2 - 600 - n^2 - 4n$$

$$P(n) = -4n^2 + 196n - 600$$

First derivative: $P'(n) = -8n + 196$

Set first derivative equal to zero:

$$-8n + 196 = 0$$

$$-8n = -196$$

$$n = 24.5$$

To determine maximum or minimum point, find the second derivative:

$$P''(n) = -8$$

When the second derivative is negative, the function is at a maximum.

Answer

24.5 million

4a.   Hondo sells bikes at a price of $150 - 5n$ each. If he sells n bikes and the total cost for these bikes is $30n + 100$, how many bikes should he sell to maximize profits?

4b.   The Jax Corporation manufactures television sets. If it costs $n^2 + 40n - 54$ dollars to manufacture a total of n sets and it can sell each set for $(49/50)n + 140$ dollars, how many sets should it manufacture to maximize profits?

Example 5

Manny's sells scanners. Manny can sell all 60 of his scanners at $30 each. For each $5 increase, he loses one sale. By how much can he raise the price to maximize sales?

Analysis

Let x = the number of $5 increases.

Let $I(x)$ = the income = number of scanners sold times the new price per scanner.

Work

number of lost sales   number of $5 increases

$$I(x) = (60 - x)(30 + 5x)$$

$$I(x) = 1{,}800 + 300x - 30x - 5x^2$$

$$I(x) = -5x^2 + 270x + 1{,}800$$

Take the first derivative: $I'(x) = -10x + 270$

Set the derivative equal to zero:

$$-10x + 270 = 0$$

$$-10x = -270$$

$$x = 27$$

Take the second derivative and test for maximum or minimum:

$$I''(x) = -10$$

A negative second derivative indicates a maximum point. To maximize profits, Manny can raise his price 27 times, each time at a $5 increase: $27 \times \$5 = \$135$.

Answer

$135

5a.   The Greenhouse sells houseplants. It normally sells 41 corn plants a week at $50 a plant. If the owner decides to increase the price by $2 per plant, she will lose one sale for each $2 increase. How much of an increase in price would maximize profits?

Exercises

5b.   The Comfy Chair Store usually sells 24 of its recliners per week at $108 each. The store manager decides to have a sale. For each drop of $3 one more chair is sold. How much of a drop in price would maximize income?

Helena, an Olympian athlete, runs 468 meters in 12 seconds. Her velocity is measured by the formula:

$$\text{Velocity} = \frac{\text{Distance}}{\text{Time}}$$

13.6 Distance and Velocity

Let's just substitute these numbers:

$$\text{Velocity} = \frac{468 \text{ meters}}{12 \text{ seconds}} = 39 \text{ meters/second}$$

39 meters/second represents Helena's *average speed* over 468 meters, $\Delta s/\Delta t$, where Δs represents the change in distance and Δt represents the change in time. Helena's speed at any particular instant may be somewhat different from the average. As Δt approaches zero, this average speed approaches her *instantaneous velocity* at any particular moment. To find the instantaneous speed, we first need to know how the distance varies with time, $s = f(t)$. Then we'll make the time interval, Δt, smaller and smaller until we get the instantaneous velocity, ds/dt.

$$v = \frac{ds}{dt} = \lim_{\Delta t \to 0} \frac{\Delta s}{\Delta t}$$

We just differentiated the distance, s, with respect to time, t, to obtain the instantaneous velocity.

Example 1

A racing car travels a distance of $s(t) = 12.7t^3 - 143t^2 + 135t$ miles, where t represents the time in hours. Find the car's velocity after 8 hours.

Analysis

The velocity is the derivative of distance with respect to time: Let $t = 8$.

Work

$$s(t) = 12.7t^3 - 143t^2 + 135t$$

Take the derivative of s with respect to time:

$$v(t) = \frac{ds}{dt} = 38.1t^2 - 286t + 135$$

$t = 8$:
$$v(t) = 38.1(8)^2 - 286(8) + 135$$
$$v(t) = 2{,}438.4 - 2{,}288 + 135 = 285.4$$

Answer

285.4 miles per hour

Exercises

1a. Julia walks a distance of $s(t) = 0.3t^3 - 0.2t^2 + 1$ miles, where t = the number of hours. Find Julia's speed after two hours.

1b. A race car travels a distance of $s(t) = 5t^3 - 4t^2 + 6t - 4$ miles, where t = the time in hours. Find the race car's speed after three hours.

Motion Affected by Gravity

Suppose an object falls to the ground. In this case, the only force exerted on the object is the pull of gravity, $16t^2$, where t = seconds. The distance the object falls, $s(t)$, is only affected by the force of gravity:

$$s(t) = 16t^2$$

Example 2

A rock is dropped off a cliff 400 feet high.

a. How many seconds does it take for the rock to hit the ground?

b. What is the velocity when it hits the ground?

Analysis

The rock is *not* projected downward. Therefore, the only force acting on the stone is the force of gravity, $16t^2$. The distance in feet is given by the formula $s(t) = 16t^2$, where t is measured in seconds, so we'll set $16t^2$ equal to the height of the cliff, 400 feet, and then solve for t.

Work

Distance: $\qquad\qquad\qquad\qquad\quad s(t) = 16t^2$

$s(t) = 400$: $\qquad\qquad\qquad\qquad 400 = 16t^2$

$\qquad\qquad\qquad\qquad\qquad\qquad 25 = t^2$

$\qquad\qquad\qquad\qquad\qquad\qquad\quad 5 = t$

Differentiate to find the velocity: $\quad v(t) = \dfrac{ds}{dt} = 32t$

$t = 5$: $\qquad\qquad\qquad\qquad\qquad v(t) = \dfrac{ds}{dt} = 32(5)$

$\qquad\qquad\qquad\qquad\qquad\qquad\qquad = 160 \text{ ft/sec}$

Answer

    a.   It takes 5 seconds for the rock to hit the ground.

    b.   The rock hits the ground at a speed of 160 ft/sec.

2a.   A ball is dropped from a window 144 feet high. How long does it take to hit the ground? What is the ball's velocity at the moment of impact?

2b.   A pen is accidentally dropped from an airplane 6,400 feet high. How long does it take to hit the ground? What is the velocity when it hits the ground?

When an object is projected upward, the height, $s(t)$, the object attains at any specific time, t, is the result of subtracting the pull of gravity downward ($16t^2$) from the initial velocity upward (feet per second, v_0):

$$s(t) = v_0 t - 16t^2$$

Exercises

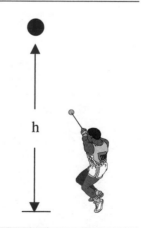

h

Example 3

Suppose that an athlete throws his hammer upward with an initial velocity of 76 ft/sec.

    a.   How high will the hammer reach after 3 seconds?

b. What is its velocity after 3 seconds?

Analysis

Let $s(t)$ = height.

Let t = time, in seconds.

Let v_0 = the initial velocity.

Use the formula for height, $s(t) = v_0 t - 16t^2$.

Let $v(t)$ = the velocity after t seconds.

Work

$$s(t) = v_0 t - 16t^2$$

$t = 3$, $v_0 = 76$: $s(3) = 76(3) - 16(3)^2$

$$s(t) = 228 - 144 = 84 \text{ feet}$$

$$v(t) = \frac{ds}{dt} = v_0 - 32t$$

$v_0 = 76$, $t = 3$: $v(3) = 76 - 32(3) = 76 - 96$
$$= -20 \, \text{ft/sec}$$

Answer

a. 84 feet

b. -20 ft/sec

Exercises

3a. Hans throws a baseball up with an initial velocity upward of 82 feet per second. How high will the ball reach in two seconds? What is its velocity in two seconds?

3b. A rocket is shot up with an initial velocity upward of 2,000 feet per second. How many feet high will it be in five seconds? What is the rocket's velocity in five seconds?

If an object is forcefully projected downward with an initial velocity, v_0, then we obtain the distance by adding the force of gravity to the initial velocity:

$$s(t) = v_0 t + 16t^2$$

Example 4

Kendra throws a ball down from the top of the Sears Tower. She throws the ball down with an initial velocity of 78 ft/sec.

a. How many feet will the ball drop after four seconds?

b. Find its velocity after four seconds.

Analysis

Use the distance formula above and substitute 4 for t.

Work

$$s(t) = v_0 t + 16t^2$$

$t = 4, v_0 = 78$: 　$s(t) = 78(4) + 16(4)^2 = 312 + 256$
$$= 568 \text{ feet}$$

$$v(t) = \frac{ds}{dt} = v_0 + 32t$$

$t = 4, v_0 = 78$: 　$v(t) = 78 + 32(4) = 78 + 128$
$$= 206 \text{ ft/sec}$$

Answer

a. 568 feet

b. 206 ft/sec

4a. An object is thrown down from an airplane with an initial velocity downward of 120 feet per second. How far will it drop in six seconds? What is its velocity in six seconds?

Exercises

4b. Gingy throws a ball down from his apartment with an initial velocity of 93 feet per second. How far will the ball fall in two seconds? What is the ball's velocity in two seconds?

Whenever we throw a ball up into the air, our initial velocity impels the ball upward while the force of gravity pulls the ball back to the ground. At a certain point, the velocity drops to zero, the ball reverses direction and falls back to the ground. The ball reaches its maximum height at this point.

Let's return to the formula in Example 3, which lets us determine the height attained by an object projected upward:

$$s(t) = v_0 t - 16t^2$$

If we take the derivative of s with respect to t, we obtain the velocity. Now, the object reaches its maximum height when the velocity is zero, so all we have to do to find out exactly when this occurs is to set the derivative equal to zero and then find t.

Example 5

An object is projected upward with an initial velocity of 320 feet per second.

    a.   After how many seconds does the object reach its maximum height?

    b.   What is the maximum height?

Analysis

Use the distance formula for an object projected upward. Find the derivative and set it equal to zero. Then find the time, t, in seconds, and substitute into the original distance function.

Work

Height formula:	$s(t) = v_0 t - 16t^2$
$v_0 = 320$:	$s(t) = 320t - 16t^2$

Find the derivative
of s with respect to t: $\quad v(t) = \dfrac{ds}{dt} = 320 - 32t$

$$v(t) = 320 - 32t$$

$v = 0$ at maximum height: $\qquad 0 = 320 - 32t$

$$32t = 320$$

$$t = 10$$

The object reaches a maximum height in 10 seconds, so let's go back to our original equation for height, $s(t) = 320t - 16t^2$, and substitute 10 for t.

$$s(t) = 320t - 16t^2$$

$t = 10$: $\qquad s(10) = 320(10) - 16(10)^2$

$$s(10) = 3{,}200 - 16(100)$$

$$s(10) = 3{,}200 - 1{,}600$$

$$s(10) = 1{,}600$$

Answer

    a.   10 seconds

    b.   1,600 feet

Exercises

5a.   An object is projected upward with an initial velocity of 192 feet per second. In how many seconds will the object reach its maximum height? What is its maximum height?

5b. A ball is thrown up with an initial velocity of 96 feet per second. In how many seconds will the ball reach its maximum height? What is its maximum height?

What is acceleration? Let's say you're a passenger in a car and the car is moving at a constant rate. All of a sudden the driver hits the brakes or else steps on the gas. The first thing you're going to feel is that surge in your stomach, which indicates that the speed of the car is changing. That's acceleration.

Acceleration is the rate at which the velocity is changing. Mathematically, if the distance is a function of time, $s = f(t)$, then:

1. Velocity is the first derivative of the distance function, s, with respect to time, t:

$$v(t) = \frac{ds}{dt}$$

2. Acceleration is the derivative of the velocity or the second derivative of the distance function, s, with respect to time, t:

$$a(t) = \frac{dv}{dt} = \frac{d^2s}{dt^2}$$

Example 1

The distance a car moves is given by the formula $s(t) = 3t^3 - 0.12(t + 3)$ feet, where t is in seconds. (a) Find the velocity at the end of three seconds. (b) Find the acceleration at the end of three seconds.

Analysis

The first derivative will tell us the velocity, and the second derivative will give us the acceleration.

Work

Start with the distance formula:

$$s(t) = 3t^3 - 0.12(t + 3) = 3t^3 - 0.12t - 0.36$$

$$v(t) = \frac{ds}{dt} = 9t^2 - 0.12$$

$t = 3$: $v(3) = 9(3)^2 - 0.12 = 81 - 0.12$

$v(3) = 80.88 \text{ ft/sec}$

$$a(t) = \frac{d^2s}{dt^2} = 18t$$

$t = 3$: $a(3) = 18(3) = 54 \, \text{ft/sec}^2$

Answer

   a. Velocity = 80.88 ft/sec

   b. Acceleration = 54 ft/sec²

Exercises

1a. A train travels a distance of $s(t) = 20t^{3/2} + 5t - 6$ feet in t seconds. Find its velocity at four seconds. Find its acceleration at four seconds.

1b. A car moves along a road according to the formula $s(t) = 2t^4 - 3t^3 + 50$ feet, where t = seconds. Find the velocity at $t = 3$. Find the acceleration at $t = 5$.

Example 2

The distance in feet a train travels is given by the function $s(t) = (1/6)t^4 + (5/6)t^3 + 6t^2 + 8$, where t = seconds. During the trip, the train changes its acceleration. At what point in time is the acceleration 64 ft/sec²?

Analysis

Find the second derivative—the acceleration, $a(t)$—and substitute 64 for a to find t.

Work

$$s(t) = (1/6)t^4 + (5/6)t^3 + 6t^2 + 8$$

First derivative: $v(t) = s'(t) = (2/3)t^3 + (5/2)t^2 + 12t$

Second derivative: $a(t) = s''(t) = 2t^2 + 5t + 12$

Let $a = 64$: $64 = 2t^2 + 5t + 12$

$$2t^2 + 5t + 12 = 64$$

$$2t^2 + 5t - 52 = 0$$

$$(2t + 13)(t - 4) = 0$$

$$2t + 13 = 0 \qquad t - 4 = 0$$

$$2t = -13 \qquad\quad t = 4$$

$$t = -13/2$$

Reject because time
should be positive.

Answer

4 seconds

2a. The distance a car moves is given by the function $s(t) = (1/12)t^4 - (2/3)t^3 - 6t^2 + 23$ feet, where t = seconds. When is the acceleration 33 ft/sec^2?

2b. The distance an airplane flies is described by the formula $s(t) = (1/12)t^4 + (1/2)t^3 - 27t^2 + 90$ miles, where t = seconds. When is the acceleration 16 miles/sec^2?

Test

1. Find $\lim\limits_{x \to 7} \dfrac{x^2 - 49}{x - 7}$.

2. Find the derivative of $f(x) = 6x^5 - 4x^3 + 2x^2 - 9x + 7$.

3. A side of a cube is increasing at the rate of 0.3 inch per second. Find the rate of increase in the volume when a side is 6 inches. $V = s^3$.

4. A restaurant determines that if they have n number of patrons, their profit is $P(n) = 300 - 50n + 3n^2$ dollars. Find the rate of profit when the restaurant serves 120 patrons.

5. The growth of mold is given by the function $M(t) = 500 + 3t^2 - 40t$ milligrams, where t = time, measured in hours. Find the *rate of growth* of bacteria when $t = 7$ hours.

6. Find the slope(s) to the curve of the graph of the function $y = 4x^2 + 2x - 3$ at the point $x = 5$.

7. Find the slope to the curve of the graph of the function $y = 2x^2 - 9x + 7$ at the point $y = -2$.

8. Find the second derivative of $f(x) = 7x^4 + 3x^3 - 5x - 4$.

9. Find the second derivative of $y = (2x^3 - 4)^2$.

10. $P(t) = 2t^3 + 3t^2 - 4t + 20$ represents the population of Hamden Township at the present time. $P'(t)$ is the rate of growth of the population. $P''(t)$ represents the change in the rate of growth of the population and t represents years. Find the change in the rate of growth of the population in four years.

11. Karl has 800 feet of fencing. He wants to enclose a rectangular piece of property, but he only needs to use fencing for three sides because he will use a wall for

the fourth side. What are the dimensions of the largest area he can enclose?

12. Julissa is going to construct an open box out of an 18-inch by 18-inch piece of cardboard by removing a square from each corner. What are the dimensions of each corner so as to maximize the volume?

13. Max sells jeans at a price of $202 - 2x$ each. If he sells x jeans and the total cost for these jeans is $50x - 100$, how many jeans should he sell to maximize profits?

14. Ace Electronics normally sells 20 cell phones a week at $200 per phone. If Ace decides to drop its price by $5, it increases its sales by one phone. How much should Ace drop its price to maximize sales?

15. A train travels a distance of $s(t) = 7t^3 - 2t^2 + 9t + 7$ miles, where $t =$ the time in hours. Find the train's rate of speed after four hours.

16. Julio throws a ball up with an initial velocity of 100 feet per second. How high will the ball reach in five seconds?

17. An object is thrown down a cliff with an initial velocity downward of 76 feet per second. How far will it drop in six seconds?

18. An object is projected upward with an initial velocity of 160 feet per second. What is its maximum height?

19. A ball is dropped from a window 1,024 feet high. What is the ball's velocity at the moment of impact?

20. A car moves along a road according to the formula $s(t) = 3t^4 + 2t^3 - 4t^2 + 20$ feet, where $t =$ seconds. (a) Find the velocity at $t = 4$. (b) Find the acceleration at $t = 3$.

14
Integration

Technically, integration is the reverse of differentiation. Graphically, integration is the area under a curve that represents a continuous function.

We integrate when we know the derivative and we wish to get back to the original function. The terms *antiderivative* and *integral* are used interchangeably. Whereas we use the symbols *dy/dx* or *f'(x)* in differentiation, we use Leibniz's symbol ∫, which represents the sum of *f(x) dx* and which is used to indicate integration.

Let's check our original statement—namely that integration is the reverse of differentiation. After we differentiate a function and integrate the derivative, we should arrive back at the original function (minus our constant, *k*).

Basic Function   *Derivative*

$$y = f(x) = x^n \qquad y' = f'(x) = nx^{n-1}$$

What can we do to the derivative, $y' = nx^{n-1}$, to change it back to our original function, $y = x^n$? If we add a 1 to the exponent and then divide by the new exponent, we arrive back at our original function.

Indefinite Integral

$$\int nx^{n-1}dx = \frac{nx^{(n-1)+1=n}}{n} = x^n + k_1$$

The integral is "indefinite" because the limits of integration are not specified.

We add a constant term, k_1, because the original function might have included a constant, but, at this point, we need more information to determine its value. For example, if the original function was either $f(x) = x^n$ or $f(x) = x^n + k_2$, the derivative, $f'(x) = nx^{n-1}$, is still the same. So it's more accurate when we reconstruct the original function to say that $f(x) = x^n + k_1$ (in this example, k_1 could have been zero or k_2).

Example 1

Evaluate $\int x^4 dx$.

Analysis

We'll get back to our original function by adding 1 to the exponent and then dividing by the new exponent.

Work

$$\int x^4 dx = \frac{x^5}{5} + k$$

Check

If $f(x) = x^5/5 + k$, then $f'(x) = x^4$, which is just the result we want!

Answer

$$\frac{x^5}{5} + k$$

Exercises

Evaluate these integrals.

1a. $\int x^3 dx$ 1b. $\int x^7 dx$

The integral of the product of a constant and a function is equal to the product of the constant and the integral of that function.

$$\int kf(x)dx = k\int f(x)dx$$

Example 2

Evaluate $\int 7x^2 dx$.

Work

$$\int 7x^2 dx = 7\int x^2 dx = \frac{7x^3}{3} + k$$

Answer

$$\frac{7x^3}{3} + k$$

Evaluate these integrals.

2a. $\int 5x^4 dx$ 2b. $\int 7x^3 dx$

The integral of a sum is the sum of the integrals.

$$\int \{f(x) + g(x)\}dx = \int f(x)dx + \int g(x)dx$$

Example 3

Evaluate $\int (3x^3 + 2x^2 - 3x + 9)dx$.

Analysis

Integrate each term separately.

Work

$$\int (3x^3 + 2x^2 - 3x + 9)dx$$

$$= \int 3x^3 dx + \int 2x^2 dx - \int 3xdx + \int 9dx$$

$$= 3\int x^3 dx + 2\int x^2 dx - 3\int xdx + 9\int dx$$

$$= \frac{3}{4}x^4 + \frac{2}{3}x^3 - \frac{3}{2}x^2 + 9x + k$$

Answer

$$\frac{3}{4}x^4 + \frac{2}{3}x^3 - \frac{3}{2}x^2 + 9x + k$$

Evaluate these integrals.

3a. $\int (4x^4 + 2x^3 - x^2 + 6x - 2)dx$

3b. $\int (5x^3 - 4x^2 + 9x - 2)dx$

There are no special rules for integration of products or quotients. Try to multiply or divide first and then integrate.

Example 4

Evaluate $\int \{(2x^2 + 3)(x - 5)\}dx$.

Work

$$\int \{(2x^2 + 3)(x - 5)\}dx = \int (2x^3 - 10x^2 + 3x - 15)dx$$

$$= \frac{x^4}{2} - \frac{10x^3}{3} + \frac{3x^2}{2} - 15x + k$$

Answer

$$\frac{x^4}{2} - \frac{10x^3}{3} + \frac{3x^2}{2} - 15x + k$$

Evaluate these integrals.

4a. $\int \{(2x + 3)(x - 4)\}dx$ 4b. $\int (x^2 - 5)(x - 2)dx$

Example 5

Evaluate $\int \sqrt{x}dx$.

Work

$$\int \sqrt{x}dx = \int x^{1/2}dx = \frac{2x^{3/2}}{3} + k$$

Answer

$$\frac{2x^{3/2}}{3} + k$$

Evaluate these integrals.

5a. $\int \sqrt[3]{x}\,dx$ 5b. $\int \sqrt[5]{x}\,dx$

When the limits of the integral are specified (as, for example, in $\int_a^b f(x)\,dx$), the integral is called a definite integral, and we can get an exact answer.

Exercises

Definite Integral

Example 6

Evaluate $\int_1^4 x^3\,dx$.

Analysis

We want to integrate between the bounds of $x = 1$ and $x = 4$.

Work

$$\int_1^4 x^3\,dx = \frac{1}{4}x^4 + k \Big]_1^4 = \left\{\frac{1}{4}(4)^4 + k\right\} - \left\{\frac{1}{4}(1)^4 + k\right\}$$

$$= \{64 + k\} - \left\{\frac{1}{4} + k\right\} = 63\frac{3}{4}$$

From our result, we see that, when we integrate within given bounds, the constant, k, drops out. From now on, we can eliminate k in these sorts of calculations.

Answer

$63\dfrac{3}{4}$

Evaluate these integrals.

6a. $\int_1^3 x^2\,dx$ 6b. $\int_2^4 x^3\,dx$

Exercises

Example 7

Evaluate $\int_{-1}^{3} 6x^4 dx$.

Work

$$\int_{-1}^{3} 6x^4 dx = 6\int_{-1}^{3} x^4 dx = \frac{6x^5}{5}\Big]_{-1}^{3} = \left\{\frac{6(3)^5}{5}\right\} - \left\{\frac{6(-1)^5}{5}\right\}$$

$$= \frac{6(243)}{5} - \frac{6(-1)}{5}$$

$$= 291\frac{3}{5} + 1\frac{1}{5} = 292\frac{4}{5}$$

Answer

$$292\frac{4}{5}$$

Exercises

Evaluate these integrals.

7a. $\int_{0}^{5} 4x^3 dx$ 7b. $\int_{3}^{6} 5x^2 dx$

Example 8

Evaluate $\int_{2}^{3}(2x^3 + 4x^2 - x + 6)dx$.

Work

$$\int_{2}^{3}(2x^3 + 4x^2 - x + 6)dx$$

$$= \left(\frac{x^4}{2} + \frac{4x^3}{3} - \frac{x^2}{2} + 6x\right)\Big]_{2}^{3}$$

$$= \left\{\frac{3^4}{2} + \frac{4(3)^3}{3} - \frac{3^2}{2} + 6(3)\right\} - \left\{\frac{2^4}{2} + \frac{4(2)^3}{3} - \frac{2^2}{2} + 6(2)\right\}$$

$$= \left\{40\frac{1}{2} + 36 - 4\frac{1}{2} + 18\right\} - \left\{8 + 10\frac{2}{3} - 2 + 12\right\}$$

$$= 90 - 28\frac{2}{3} = 61\frac{1}{3}$$

Answer

$$61\frac{1}{3}$$

Evaluate these integrals.

8a. $\int_1^2 (5x^3 - 2x^2 + x - 5)dx$

8b. $\int_2^4 (4x^2 + 8x - 3)dx$

Prior to the 17th century, the methods used for finding the area under a curve yielded fairly inaccurate results. All sorts of mathematical gimmicks were used to determine areas. Toward the end of that century, both the German mathematician Wilhelm Gottfried Leibniz and the English mathematician Isaac Newton solved the problem independently, and they both developed the same method for finding the area under a curve $y = f(x)$ between $x = a$ and $x = b$.

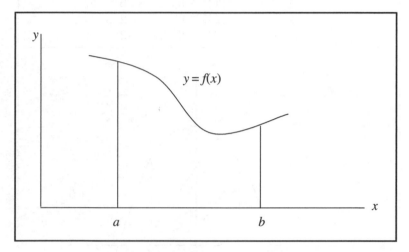

The **Fundamental Theorem of Calculus** states:

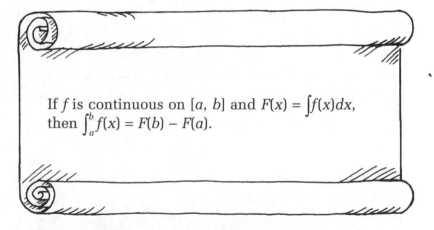

If f is continuous on $[a, b]$ and $F(x) = \int f(x)dx$, then $\int_a^b f(x) = F(b) - F(a)$.

What does this mean? It means that, in order to find the area under the curve, we must first integrate the function, substitute the upper and lower limits into the new function, and then subtract the two limits.

Since the time of Leibniz and Newton, lots of other applications of the Fundamental Theorem—ranging from economics to biology—have been developed.

Example 1

Find the area bounded by the graph of the function $y = -x^2 + 6$, the x-axis, and the lines $x = -1$ and $x = 1$.

x	$-x^2 + 6$	y
-3	$-(-3)^2 + 6$	-3
-2	$-(-2)^2 + 6$	2
-1	$-(-1)^2 + 6$	5
0	$-(0)^2 + 6$	6
1	$-(1)^2 + 6$	5
2	$-(2)^2 + 6$	2
3	$-(3)^2 + 6$	-3

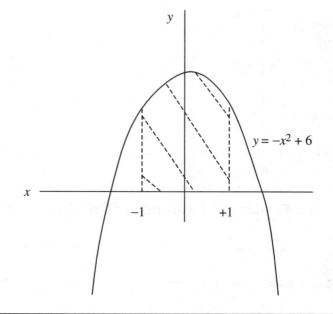

$y = -x^2 + 6$

Analysis

Integrate the function with limits of 1 and -1.

Work

$$\text{Area} = \int_{-1}^{1} \left(-x^2 + 6\right)dx$$

$$= -\frac{x^3}{3} + 6x \Bigg]_{-1}^{1} = \left\{-\frac{(1)^3}{3} + 6(1)\right\} - \left\{-\frac{(-1)^3}{3} + 6(-1)\right\}$$

$$= 5\frac{2}{3} + 5\frac{2}{3} = 11\frac{1}{3}$$

Answer

$$11\frac{1}{3}$$

1a. Find the region bounded by $f(x) = -2x^2 + 3x + 9$ and the lines $x = 0$ and $x = 2$.

1b. Find the area bounded by $y = x^2 + 3x + 1$ and the lines $x = 3$ and $x = 2$.

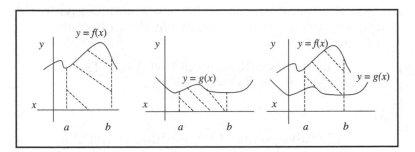

If an area is bounded by $f(x)$ and $g(x)$ and the vertical lines $x = a$ and $x = b$ and both functions are continuous on the interval $a \le x \le b$ and $f(x) \ge g(x)$, then the area between f and $g =$

$$\int_a^b f(x)dx - \int_a^b g(x)dx$$

$$= \int_a^b \{f(x) - g(x)\}dx$$

Example 2

Find the area bounded by the graph of the function $f(x) = x^2 - x - 6$ and the x-axis.

x	$x^2 - x - 6$	$f(x)$
−3	$(-3)^2 - (-3) - 6$	6
−2	$(-2)^2 - (-2) - 6$	0
−1	$(-1)^2 - (-1) - 6$	−4
0	$(0)^2 - (0) - 6$	−6
1	$(1)^2 - (1) - 6$	−6
2	$(2)^2 - (2) - 6$	−4
3	$(3)^2 - (3) - 6$	0

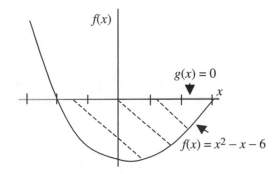

Analysis

Both the x-axis $\{g(x) = 0\}$ and f are continuous in the bounded interval. First, find the points of intersection. Then, since $g \geq f$ over all points in the interval, subtract f from g and integrate the difference of the two functions, with limits indicated by the abscissa at the points of intersection.

The x-axis is the function $g(x) = 0$. The second function is $f(x) = x^2 - x - 6$.

Work

Find the points of intersection of the graphs $g(x) = 0$ and $f(x) = x^2 - x - 6$ by finding points that satisfy both equations simultaneously:

$$x^2 - x - 6 = 0$$

Factor:
$$(x - 3)(x + 2) = 0$$

$$x - 3 = 0 \qquad x + 2 = 0$$

$$x = 3 \qquad x = -2$$

The area we want is below the line $g(x) = 0$ but above the curve $f(x) = x^2 - x - 6$, so we'll subtract:

$$g(x) - f(x) = (0) - (x^2 - x - 6) = -x^2 + x + 6$$

We only want the section of this area between $x = -2$ and $x = 3$, so we'll integrate with limits of 3 and −2.

$$A = \int_{-2}^{3} \left(-x^2 + x + 6 \right) dx$$

$$A = -\frac{1}{3}x^3 + \frac{1}{2}x^2 + 6x \Big]_{-2}^{3}$$

$$= \left\{ -\frac{1}{3}(3)^3 + \frac{1}{2}(3)^2 + 6(3) \right\}$$

$$- \left\{ -\frac{1}{3}(-2)^3 + \frac{1}{2}(-2)^2 + 6(-2) \right\}$$

$$= \left\{ 13\frac{1}{2} \right\} - \left\{ -7\frac{1}{3} \right\} = 20\frac{5}{6}$$

Answer

$$20\frac{5}{6}$$

2a. Find the region bounded by the function $y = x^2 - 4x + 3$ and the x-axis.

2b. Find the area bounded by $f(x) = x^2 + 1x - 2$ and the x-axis.

Example 3

Find the area bounded by the curves $g(x) = x^2 - 2x$ and $f(x) = 4 - x^2$.

x	$x^2 - 2x$	$g(x)$		x	$4 - x^2$	$f(x)$
−3	$(-3)^2 - 2(-3)$	15		−3	$4 - (-3)^2$	−5
−2	$(-2)^2 - 2(-2)$	8		−2	$4 - (-2)^2$	0
−1	$(-1)^2 - 2(-1)$	3		−1	$4 - (-1)^2$	3
0	$(0)^2 - 2(0)$	0		0	$4 - (0)^2$	4
1	$(1)^2 - 2(1)$	−1		1	$4 - (1)^2$	3
2	$(2)^2 - 2(2)$	0		2	$4 - (2)^2$	0
3	$(3)^2 - 2(3)$	3		3	$4 - (3)^2$	−5

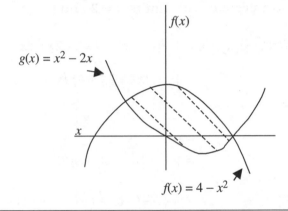

Analysis

Find the points of intersection by setting the two equations equal. When we have the points of intersection, let's see which function is greater between those two points. Then find the definite integral of the difference of the two functions, with limits indicated by the abscissa at the points of intersection.

Work

To find the points of intersection, set the two functions equal:

$$x^2 - 2x = 4 - x^2$$
$$2x^2 - 2x - 4 = 0$$
$$x^2 - x - 2 = 0$$

Factor: $(x - 2)(x + 1) = 0$

$$x - 2 = 0 \qquad x + 1 = 0$$

$$x = 2 \qquad x = -1$$

At the points of intersection, therefore, $x = 2$ and $x = -1$.

If we want to find the corresponding ordinates, simply take either of the functions and substitute the abscissa values:

$$y = 4 - x^2$$

$x = 2$: $\quad y = 4 - (2)^2 = 0$

$x = -1$: $\quad y = 4 - (-1)^2 = 3$

The points of intersection are, therefore, (2, 0) and (−1, 3).

Now let's look at the preceding table to determine which function is greater between the two points of intersection. At all these points, $f(x) > g(x)$, so we can now subtract $g(x)$ from $f(x)$ and integrate with limits $x = 2$ and $x = -1$.

$$\int_{-1}^{2} \{f(x) - g(x)\}dx = \int_{-1}^{2} \{(4 - x^2) - (x^2 - 2x)\}dx$$

$$= \int_{-1}^{2} (-2x^2 + 2x + 4)dx$$

$$= -\frac{2}{3}x^3 + x^2 + 4x \bigg]_{-1}^{2}$$

$$= \left\{-\frac{2}{3}(2)^3 + (2)^2 + 4(2)\right\}$$

$$\quad - \left\{-\frac{2}{3}(-1)^3 + (-1)^2 + 4(-1)\right\}$$

$$= \left\{-\frac{2}{3}(8) + 4 + 8\right\} - \left\{+\frac{2}{3} + 1 - 4\right\}$$

$$= \left\{-\frac{16}{3} + 12\right\} - \left\{1\frac{2}{3} - 4\right\}$$

$$= 6\frac{2}{3} + 2\frac{1}{3} = 9$$

Answer

9

Exercises

3a. Find the region bounded by the curves $f(x) = 2x^2 - 2x + 5$ and $g(x) = x^2 - x + 25$.

3b. Find the area bounded by the graph of the parabola $y = -x^2 - x + 2$ and a second parabola, $y = x^2 + x + 2$.

Example 4

Find the area bounded by $f(x) = x - 5$, $g(x) = -2x^2 + 3$, $x = -1$ and $x = 1$.

x	$x - 5$	$f(x)$		x	$-2x^2 + 3$	$g(x)$
-3	$-3 - 5$	-8		-3	$-2(-3)^2 + 3$	-15
-2	$-2 - 5$	-7		-2	$-2(-2)^2 + 3$	-5
-1	$-1 - 5$	-6		-1	$-2(-1)^2 + 3$	$+1$
0	$0 - 5$	-5		0	$-2(0)^2 + 3$	$+3$
$+1$	$+1 - 5$	-4		1	$-2(1)^2 + 3$	$+1$
$+2$	$+2 - 5$	-3		2	$-2(2)^2 + 3$	-5
$+3$	$+3 - 5$	-2		3	$-2(3)^2 + 3$	-15

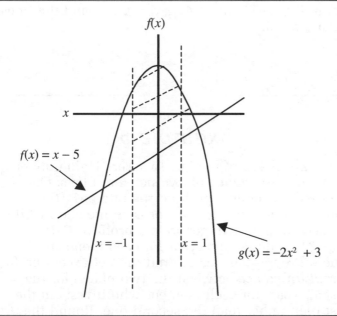

Analysis

We need to find the area between the lines $x = -1$ and $x = 1$ but also below $g(x)$ and above $f(x)$. We must subtract the functions and then integrate with limits of 1 and -1.

Work

$$\text{Area} = \int_{-1}^{1} \{g(x) - f(x)\}dx = \int_{-1}^{1} \{-2x^2 + 3\} - \{x - 5\}dx$$

$$= \int_{-1}^{1} \{-2x^2 - x + 8\}dx = \left[-\frac{2}{3}x^3 - \frac{1}{2}x^2 + 8x \right]_{-1}^{1}$$

$$= \left\{ -\frac{2}{3}(1)^3 - \frac{1}{2}(1)^2 + 8(1) \right\}$$

$$- \left\{ -\frac{2}{3}(-1)^3 - \frac{1}{2}(-1)^2 + 8(-1) \right\}$$

$$= \left\{6\frac{5}{6}\right\} - \left\{-7\frac{5}{6}\right\} = 14\frac{2}{3}$$

Answer

$$14\frac{2}{3}$$

Exercises

4a. Find the region bounded by the graph of the function $f(x) = -2x^2 + 6x + 8$, the x-axis, and the lines $x = 0$ and $x = 3$.

4b. Find the area bounded by the graph of the parabola $f(x) = -x^2 + 3x + 4$, $g(x) = x + 1$, and the lines $x = 1$ and $x = 3$.

Example 5

Local 459 of the Steelworkers Union offers its members a choice of two pension plans. One plan will generate a profit at the rate of $P_1(t) = 400 + 0.6t$ dollars per year for each member, while the second plan will generate a profit of $P_2(t) = 5t^2 - 312.8$ dollars per year for each member, where t represents years. Find the net excess profit (*profit difference* between the two plans) for the first 12 years for each member who invests in the first plan rather than the second one. Round the answer to the nearest cent.

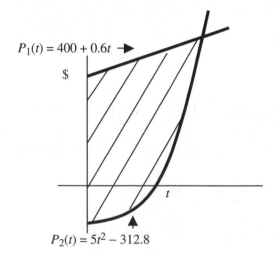

$P_1(t) = 400 + 0.6t$

$\$$

t

$P_2(t) = 5t^2 - 312.8$

Analysis

To see the excess profit, draw both functions on the same graph. To determine the profit, integrate the difference of the two functions.

Work

$$P_1(t) = 400 + 0.6t$$

t	$400 + 0.6t$	$P_1(t)$
0	400 + 0.6(0)	400.0
3	400 + 0.6(3)	401.8
6	400 + 0.6(6)	403.6
9	400 + 0.6(9)	405.4
12	400 + 0.6(12)	407.2
15	400 + 0.6(15)	409.0

$$P_2(t) = 5t^2 - 312.8$$

t	$5t^2 - 312.8$	$P_1(t)$
0	$5(0)^2 - 312.8$	-312.8
3	$5(3)^2 - 312.8$	-267.8
6	$5(6)^2 - 312.8$	-132.8
9	$5(9)^2 - 312.8$	92.2
12	$5(12)^2 - 312.8$	407.2
15	$5(15)^2 - 312.8$	812.2

Net excess profit:

$$NEP = \int_0^{12} (400 + 0.6t) - (5t^2 - 312.8)dt$$

$$= \int_0^{12} (712.8 - 5t^2 + 0.6t)dt$$

$$= \int_0^{12} (-5t^2 + 0.6t + 712.8)dt$$

$$NEP = \left[\frac{-5t^3}{3} + 0.3t^2 + 712.8t \right]_0^{12}$$

$$= \{-1.67(12)^3 + 0.3(12)^2 + 712.8(12)\} - \{0\}$$

$$= \{-2{,}885.76 + 43.2 + 8{,}553.6\}$$

$$= 5{,}711.04$$

Answer

$5,711.04

5a. The Sentex Corporation is considering two business plans. One plan will generate a profit of $P_1(t) = (-8t + 500)$ millions of dollars per year. A second business plan will generate a profit of $P_2(t) = (4t^2 - 300)$ millions of dollars per year, where t represents years. Find the net excess profit (profit difference between the two plans) for the first five years ($t = 0$ through $t = 5$).

Exercises

5b. The New Orleans Cruisers, a basketball team, is considering two marketing plans. One plan will produce a profit of $P_1(t) = (6t^2 - 18)$ millions of dollars per year. A second plan will produce a profit of $P_2(t) = (6 + 0.3t)$ millions of dollars per year, where t represents years. Find the net excess profit (profit difference between the two plans) for the first two years ($t = 0$ through $t = 2$).

Example 6

The cost of operating a sewing machine is $C(t) = 30 + 3t^3$ and the gross income is $GI(t) = \$400 - t^2$, where t represents days. Find the net income, NI (gross income minus cost), between the second and fourth days of operation of the sewing machine. Round your answer to the nearest cent.

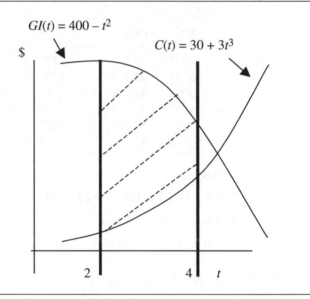

$GI(t) = 400 - t^2$

$C(t) = 30 + 3t^3$

Analysis

We'll graph both equations on the same pair of axes. Then we'll determine the integral for the area in between the two curves with limits of 2 and 4.

Work

t	$30 + 3t^3$	$C(t)$
1	$30 + 3(1)^3$	33
2	$30 + 3(2)^3$	54
3	$30 + 3(3)^3$	111
4	$30 + 3(4)^3$	222
5	$30 + 3(5)^3$	405

t	$400 - t^2$	$GI(t)$
1	$400 - (1)^2$	399
2	$400 - (2)^2$	396
3	$400 - (3)^2$	391
4	$400 - (4)^2$	384
5	$400 - (5)^2$	375

$C(t) = 30 + 3t^3$

$GI(t) = 400 - t^2$

Net income = Gross income − Cost:

$$NI = \int_{2}^{4}\{(400 - t^2) - (30 + 3t^3)\}dt$$

$$NI = \int_{2}^{4}(370 - 3t^3 - t^2)dt$$

$$= \int_{2}^{4}(-3t^3 - t^2 + 370)dt$$

$$NI = \left[\frac{-3t^4}{4} - \frac{t^3}{3} + 370t\right]_{2}^{4}$$

$$NI = \left\{\frac{-3(4)^4}{4} - \frac{4^3}{3} + 370(4)\right\}$$

$$-\left\{\frac{-3(2)^4}{4} - \frac{2^3}{3} + 370(2)\right\}$$

$$NI = \{-192 - 21.333 + 1,480\}$$
$$- \{-12 - 2.667 + 740\}$$

$$NI = \{1,266.667\} - \{725.333\}$$

$$NI = 541.334$$

Answer

$541.33

6a.   The annual cost of operating a gas station is $C(t) = (25 + 2t^3)$ thousands of dollars while the gross income is $GI(t) = (325 - 4t)$ thousands of dollars, where t represents years. Find the net income, NI (gross income minus cost), between the third and fifth years of operation.

6b.   The monthly cost of running a hotel is $C(t) = (3t^2 - 18)$ thousands of dollars. The monthly gross income is $GI(t) = (320 - 2t)$ thousands of dollars. Find the net income, NI (gross income minus cost) between the first and fifth months.

Exercises

14.3 Marginal Cost, Marginal Revenue

   Economists have borrowed the idea of the derivative from calculus and have applied it to economics. Thus, if $C(x)$ is the cost of producing x units, $C'(x)$ is the cost of producing that one extra unit and is called the **marginal cost**. The marginal cost is the instantaneous rate of change of C with respect to x, $C'(x)$. Therefore, if we are given the marginal cost, $C'(x)$, for producing that one extra unit, we can work backward to find the total cost of producing x items, $C(x)$.

   The **fixed cost** is the cost of just remaining in business without producing any units at all (rent, electricity, mortgage, etc.).

Example 1

Methuselah Aerobics has determined that the marginal cost for manufacturing an exercise machine is given by $C'(x) = -0.6x + 10$, where x represents the number of machines. If the fixed cost, k, is \$20, find the manufacturing cost, $C(x)$, and graph it for $10 \leq x \leq 60$.

Analysis

Start with the marginal cost, $C'(x)$, integrate to find $C(x)$, and then graph it for limits of 10 and 60.

Work

$$C'(x) = -0.6x + 10$$
$$C(x) = \int(-0.6x + 10)dx$$
$$= -0.3x^2 + 10x + k$$

$k = 20$:
$$C(x) = -0.3x^2 + 10x + 20$$

x	$C(x) = -0.3x^2 + 10x + 20$
10	90
20	100
30	50
40	−60
50	−230
60	−460

$$C(x) = -0.3x^2 + 10x + 20$$

Answer

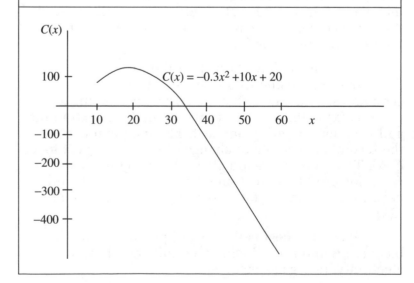

1a.   The marginal cost for reproducing a video is $C'(x) = 2 - 0.10x$ dollars, where x represents units. If the fixed cost is $30, develop the function describing the cost of producing x videos.

1b.   The marginal cost of manufacturing x pens is $C'(x) = 40 - 0.10x^2$, where x represents the number of pens. If the fixed cost is $25, find the function describing the cost of producing x pens.

Example 2

The marginal cost, $C'(x)$, of manufacturing a small engine is $(6x^2 - 50x + 700)$ dollars, where x represents the number of engines. If it costs $1,979 to manufacture the first three engines, how much would it cost to manufacture the first six engines?

Analysis

Integrate to find the cost of producing x engines. Substitute $1,979 for $C(3)$ to find the constant, k. Then substitute 6 for x and the value for k just determined.

Work

$$C'(x) = 6x^2 - 50x + 700$$

$$C(x) = \int (6x^2 - 50x + 700)dx$$

$$= 2x^3 - 25x^2 + 700x + k$$

It costs $1,979 to produce the first three motors.

$$C(3) = 2(3)^3 - 25(3)^2 + 700(3) + k$$

$$1,979 = 54 - 225 + 2,100 + k$$

$$1,979 = 1,929 + k$$

$$k = 50$$

Substitute 50 for k and 6 for x in $C(x)$:

$$C(6) = 2(6)^3 - 25(6)^2 + 700(6) + 50$$

$$= 432 - 900 + 4,200 + 50 = 3,782$$

Answer

$3,782

2a. The marginal cost for manufacturing a table is $C'(x) = \{(1/9)x^2 - (1/2)x + 2\}$ dollars. If it costs \$95 to manufacture the first three tables, how much would it cost to manufacture the first six tables?

2b. The marginal cost of manufacturing x portable radios is $C'(x) = \{-\frac{1}{2}x^2 + 5x + 2\}$ dollars. It costs \$80 to manufacture the first six radios. Find the cost of manufacturing the first 12 radios.

Example 3

JetSet Ski Corporation finds that its marginal cost function in producing skis is $C'(x) = 3 + 0.4x$. Find the increase in cost when the company raises its output from 200 units to 250 units.

Analysis

Integrate and then subtract $C(200)$ from $C(250)$.

Work

$$C'(x) = 3 + 0.4x$$

$$C(x) = \int_{200}^{250} (3 + 0.4x)dx$$

$$C(x) = 3x + 0.2x^2 \Big]_{200}^{250}$$

$$C(250) - C(200) = \left\{3(250) + 0.2(250)^2\right\}$$

$$- \left\{3(200) + 0.2(200)^2\right\}$$

$$= \{750 + 12{,}500\} - \{600 + 8{,}000\}$$

$$= \{13{,}250 - 8{,}600\}$$

$$= 4{,}650$$

Answer

$4,650

3a. The marginal cost of manufacturing x golf clubs is $C'(x) = (50 - 0.06x)$ dollars. Find the increase in cost when raising output from 30 to 60 golf clubs.

3b. The marginal cost of producing x staplers is $C'(x) = (8 + 0.03x)$ dollars. Find the cost of raising production from 300 to 500 staplers.

Marginal revenue, $R'(x)$, is the revenue earned by selling that one extra unit. It is the instantaneous rate of change of the revenue, $R(x)$, with respect to the number of units sold, x.

Example 4

The marginal revenue earned, $R'(x)$, in selling an item is $x^3 - 5x + 9$ per item where x represents the number of items. If a retailer earns $33 in selling the first two items, how much would she earn in selling the first five items?

Analysis

Integrate to find the revenue in selling x items. Find k by substituting two for x and $33 for $R(2)$. Then find $R(5)$ by substituting 5 for x and the value for k just determined.

Work

$$R'(x) = x^3 - 5x + 9$$

$$R(x) = \int (x^3 - 5x + 9)dx$$

$$= \frac{1x^4}{4} - \frac{5x^2}{2} + 9x + k$$

$33 is earned on the first two items.

$$R(2) = \frac{1(2)^4}{4} - \frac{5(2)^2}{2} + 9(2) + k$$

$$33 = 4 - 10 + 18 + k$$

$$33 = 12 + k$$

$$k = 21$$

$$R(x) = \frac{1x^4}{4} - \frac{5x^2}{2} + 9x + 21$$

$$R(5) = \frac{1(5)^4}{4} - \frac{5(5)^2}{2} + 9(5) + 21$$

$$= 156.25 - 62.50 + 45 + 21 = 159.75$$

Answer

$159.75

4a. The marginal revenue earned selling x dresses is $R'(x) = (3x^2 + 3x - 8)$ dollars. If a retailer earns $96 selling the first three dresses, how much would he earn in selling the first seven dresses?

4b. The marginal revenue earned in selling a television set is $R'(x) = (x^2 + 5x - 8)$ dollars. If a department store earns $660 in selling the first six television sets, how much would it earn selling the first nine sets?

Exercises

Example 5

HoneyPack Corporation produces a special type of printer. The company analyzed its cost structure and has discovered that its marginal revenue is $R'(x) = (9\sqrt{x} - 8)$ dollars, where $x =$ the number of printers. Find the additional revenue produced when increasing output from 100 to 400 printers.

Analysis

Integrate and find the revenue produced by selling x printers, $R(x)$. Then subtract $R(100)$ from $R(400)$.

Work

$$R'(x) = 9\sqrt{x} - 8$$

$$R(x) = \int_{100}^{400} \left(9(x)^{1/2} - 8\right)dx = 6x^{3/2} - 8x\Big]_{100}^{400}$$

$$R(400) - R(100) = \left\{6(400)^{3/2} - 8(400)\right\}$$

$$- \left\{6(100)^{3/2} - 8(100)\right\}$$

$$= \{6(8,000) - 3,200\} - \{6(1,000) - 800\}$$

$$= \{48,000 - 3,200\} - \{6,000 - 800\}$$

$$= 44,800 - 5,200$$

$$= 39,600$$

Answer

$39,600

5a. Hammers Athletics manufactures tennis rackets. The marginal revenue is $R'(x) = (3\sqrt{x} - 2)$ dollars, where $x =$ the number of rackets. Find the additional revenue produced when increasing output from 25 to 100 rackets.

5b. The Western Corporation runs cattle roundup tours for amateur cowboys. Its marginal revenue is described as $R'(x) = (5 + 4x)$ dollars, where $x =$ the number of tourists. Find the additional revenue produced when increasing the number of tourists from 500 to 900.

Let's determine the volume generated by rotating the region bounded by $f(x)$, $x = k$, and $y = 0$ about the x-axis. We'll slice up the figure into thin disks—each dx thickness.

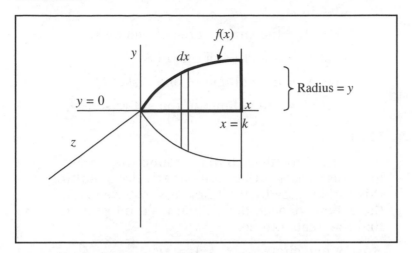

The volume of each disk is $\pi r^2 dx = \pi y^2 dx$. To obtain the total volume of the three-dimensional figure, we'll add up all the disks:

$$V = \pi \int y^2 dx$$

So far, so good. However, we can't integrate when we have y^2 under the integral sign because we have to integrate with respect to x.

So we must substitute $f(x)$ for y:

$$V = \pi \int_0^k [f(x)]^2 dx$$

Example 1

Find the volume of the solid formed by rotating the curve $y = x^2$ around the x-axis and bounded by the planes $x = 4$ and $y = 0$. Leave the answer in terms of π.

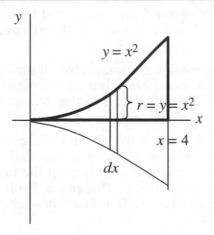

Rotation Around the x-Axis

Analysis

We'll divide the volume into various disks. Then we'll integrate to determine the volume of the full solid.

Let A_1 = the surface area of one disk.

Let r = the radius of one disk.

Let V_1 = the volume of one disk.

Let V_T = the volume of the entire solid.

Work

We'll rotate each disk around the x-axis. Each disk has a surface area of πr^2. If we multiply this surface area by the thickness, dx, we obtain the volume of each disk. Then we'll integrate to find the total volume.

Area of one circle: $\qquad A_1 = \pi r^2$

$r = y$: $\qquad\qquad\qquad A_1 = \pi y^2$

$y = x^2$ $\qquad\qquad\qquad A_1 = \pi(x^2)^2 = \pi x^4$

Volume of one disk: $\qquad V_1 = \pi x^4 dx$

Total volume of solid: $\qquad V_T = \int_0^4 \pi x^4 dx$

$$V_T = \pi\left[\frac{x^5}{5}\right]_0^4$$

$$V_T = \pi\left[\frac{4^5}{5} - \frac{0^5}{5}\right]$$

$$V_T = 204.8\pi$$

Answer

204.8π

1a. Find the volume of the solid formed by rotating the curve $y = x^3$ around the x-axis and bounded by the planes $x = 3$ and $y = 0$. Leave the answer in terms of π.

1b. Find the volume generated by rotating $y = x^4$ around the x-axis and bounded by the planes $x = 2$ and $y = 0$. Let $\pi = 3.14$ and round to the nearest whole number.

Rotation Around the y-Axis

Let's find the volume of the solid generated by rotating $y = f(x)$ around the y-axis bounded by the planes $y = k$ and $x = 0$. We'll slice up the figure into the thinnest disks possible—dy thickness. Each radius is x. Now the volume of each disk is $\pi r^2 dy = \pi x^2 dy$. We'll add up all the disks to get the total volume of the three-dimensional figure:

$$V = \pi \int x^2 dy$$

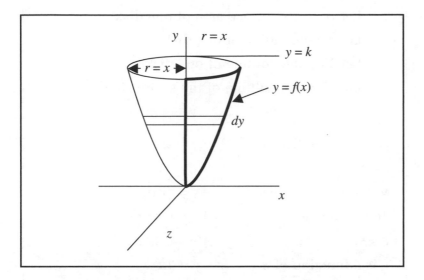

We can't integrate when we have x^2 under the integral sign because we have to integrate with respect to y. We must rearrange the function so that $x = g(y)$ and then substitute for x^2 under the integral sign:

$$V = \pi \int_0^k [g(y)]^2 dy$$

Example 2

Find the volume of the solid generated by rotating $y = 2x^2$ around the y-axis and bounded by the planes $y = 5$ and $x = 0$. Leave the answer in terms of π.

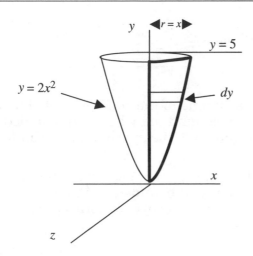

Analysis

First, find the volume of one disk. Then integrate to find the total volumes.

Let A_1 = the surface area of one disk.

Let r = the radius of one disk.

Let V_1 = the volume of one disk.

Let V_T = the volume of the entire solid.

Work

Area of one disk: $\qquad A_1 = \pi r^2$

$r = x$: $\qquad A_1 = \pi x^2$

Volume of one disk: $\qquad V_1 = \pi r^2 dy$

Total volume: $\qquad V_T = \int_0^5 \pi r^2 dy$

$r = x$: $\qquad V_T = \int_0^5 \pi x^2 dy$

y is a function of x: $\qquad y = 2x^2$

$$\sqrt{y/2} = x$$

$$x = (y/2)^{1/2}$$

$$x^2 = y/2$$

Go back to V_T: $\qquad V_T = \int_0^5 \pi x^2 dy$

Substitute $y/2$ for x^2: $\qquad V_T = \pi \int_0^5 \left[\dfrac{y}{2}\right]^2 dy$

$$V_T = \pi \int_0^5 \left[\dfrac{y^2}{4}\right] dy = \pi \left[\dfrac{y^3}{12}\right]_0^5$$

$$= \pi \left[\dfrac{5^3}{12} - \dfrac{0^3}{12}\right]$$

$$= \dfrac{125}{12}\pi$$

Answer

$$\text{Volume} = \tfrac{125}{12}\pi$$

Exercises

2a. Find the volume of the solid generated by rotating the curve $y = x^2$ around the y-axis and bounded by the planes $y = 3$ and $x = 0$. Leave your answer in terms of π.

2b. Find the volume generated by rotating $y = 3x^2$ around the y-axis and bounded by the planes $y = 2$ and $x = 0$. Leave your answer in terms of π.

Example 3

Find the volume of the solid formed by rotating $y = 9 - x^2$ about the line $y = 3$ and bounded by the planes $x = 2$ and $y = 3$. Leave the answer in terms of π.

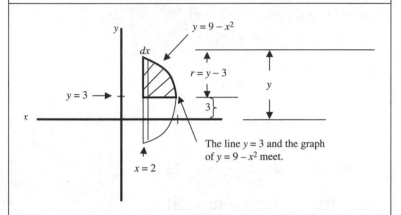

The line $y = 3$ and the graph of $y = 9 - x^2$ meet.

Analysis

Let's divide the solid into disks. We'll find the volume of one disk and then integrate from $x = 2$ to the point where the line $y = 3$ and the graph of $y = 9 - x^2$ meet.

Work

We don't know the upper limit yet, but it exists at the point where the line $y = 3$ and the graph of $y = 9 - x^2$ meet. So, let's just substitute 3 for y in the second equation:

$$y = 9 - x^2$$

$y = 3$: $\quad 3 = 9 - x^2$

$$-6 = -x^2$$

$$x = \pm\sqrt{6}$$

Reject $x = -\sqrt{6}$, so the upper limit is $x = +\sqrt{6}$.

Surface area of one disk: $\quad A_1 = \pi r^2$

$r = y - 3$: $\quad A_1 = \pi(y - 3)^2$

Volume of one disk: $\quad V_1 = \pi(y - 3)^2 dx$

To integrate, we must exchange y for an equivalent value of x, so let's go back to one of our original equations:

$$y = 9 - x^2$$
$$y - 3 = (9 - x^2) - 3$$
$$y - 3 = 6 - x^2$$

Go back to V_1: $\qquad V_1 = \pi(y - 3)^2 dx$

$y - 3 = 6 - x^2$: $\qquad V_1 = \pi(6 - x^2)^2 dx$

Integrate to find
the total volume: $\qquad V_T = \pi \int_2^{\sqrt{6}} (6 - x^2)^2 dx$

$$V_T = \pi \int_2^{\sqrt{6}} (36 - 12x^2 + x^4) dx$$

$$V_T = \pi \left[36x - 4x^3 + \frac{x^5}{5} \right]_2^{\sqrt{6}}$$

$$V_T = \pi \left[\left\{ 36\sqrt{6} - 4\left(\sqrt{6}\right)^3 + \frac{\left(\sqrt{6}\right)^5}{5} \right\} - \left\{ 36(2) - 4(2)^3 + \frac{2^5}{5} \right\} \right]$$

$$V_T = \pi \left[36\sqrt{6} - 24\sqrt{6} + \frac{36}{5}\sqrt{6} - 72 + 32 - \frac{32}{5} \right]$$

$$V_T = \pi [12\sqrt{6} + 7.2\sqrt{6} - 40 - 6.4]$$

$$V_T = [19.2\sqrt{6} - 46.4]\pi$$

Answer

$$[19.2\sqrt{6} - 46.4]\pi$$

Exercises

3a. Find the volume generated by rotating $y = 6 - x^2$ around the line $y = 2$ and bounded by the planes $y = 2$ and $x = 0$. Leave your answer in terms of π.

3b. Find the volume generated by rotating $y = 4 - x^2$ around the line $y = 1$ and bounded by the planes $y = 1$ and $x = 0$. Leave your answer in terms of π.

Test

1–8. Evaluate the following integrals.

1. $\int x^4 dx$

2. $\int 3x^2 dx$

3. $\int (4x^3 + 2x^2 - 3x + 4) dx$

4. $\int \{(x + 5)(2x - 1)\} dx$

5. $\int 6\sqrt{x} dx$

6. $\int_2^4 x^2 dx$

7. $\int_1^5 3x^2 dx$

8. $\int_0^2 (3x^4 + 5x^3 - 2x + 4) dx$

9. Find the region bounded by $f(x) = x^2 - 3x + 7$, the x axis, and the lines $x = 1$ and $x = 3$.

10. Find the region bounded by the function $y = x^2 + 6x + 8$, the x-axis, and the lines $x = 0$ and $x = 2$.

11. Find the region bounded by the curves $f(x) = x^2 + 3x - 2$ and $g(x) = 2x^2 + 1x - 10$.

12. Find the region bounded by the graph of the function $f(x) = 3x^2 - 2x + 5$, the x-axis, and the lines $x = 0$ and $x = 4$.

13. Diva Records is planning on releasing a new compact disk. One plan will produce a profit of $P_1(t) = (5t^2 - 9)$ millions of dollars per year. A second plan will produce a profit of $P_2(t) = (7 + 3t)$ millions of dollars per year, where t represents years. Find the net excess profit (profit difference between the two plans) for the first two years (from $t = 0$ through $t = 2$).

14. The annual cost of operating a supermarket is $C(t) = (4t^2 - 16)$ thousands of dollars. The annual gross income is $GI(t) = (260 - 1.3t)$ thousands of dollars. Find the net income, NI (gross income minus cost), between the third and sixth years.

15. The marginal cost for manufacturing a pen is $C'(x) = 6 - 0.20x$ dollars, where x represents units. If the fixed cost is $200, develop the function describing the cost of manufacturing x pens.

16. Find the volume of the solid formed by rotating the curve $y = x^4$ around the x-axis and bounded by the planes $x = 3$ and $y = 0$. Leave the answer in terms of π.

17. The marginal cost of manufacturing x scanners is $C'(x) = (25 - 0.007x)$ dollars. Find the increase in cost when raising output from 200 to 600.

18. The marginal revenue earned selling x cameras is $R'(x) = (2x + 4)$ dollars. If a retailer earns $820 selling the first eight cameras, how much would she earn in selling the first 12 cameras?

19. The Jax Corporation manufactures compact disks. The marginal revenue in this production is described as $R'(x) = (6x - 2)$ thousands of dollars, where $x =$ the number of *master disks*. Find the additional revenue produced when increasing the number of master disks from four to nine.

20. Find the volume generated by rotating $y = 2x^2$ around the y-axis and bounded by the planes $y = 3$ and $x = 0$. Leave your answer in terms of π.

Appendix

1.2

1a. $V = lwh$

1b. $P = 2l + 2w$

1c. $M = \frac{a+b+c+d}{4}$

1d. $s = wh$ or $s = hw$

2a. 259.2 cubic inches

2b. 27.2 feet

2c. 80

2d. $588.24

1.3

1a. $m = \frac{c-3}{2.20}$

1b. $a = P - b - c$

1c. $i = A - p$

2a. $r = \frac{C}{2\pi}$

2b. $h = \frac{V}{\pi r^2}$

1.4

1a. $C = 55°$ Celsius

1b. $C = 40°$ Celsius

2a. $F = 95°$ Fahrenheit

2b. $F = 50°$ Fahrenheit

1.5

1a. $972

1b. $1,000

2a. $2,714

2b. $4,954

3a. 4%

3b. 3 years

1.6

1a. 384.3 miles

1b. 452.2 miles

2a. 7.5 hours

2b. 58.4 mph

3a. 3:30 P.M.

3b. 8:30 P.M.

1.7

1a. 116 pounds

1b. $17,800

2a. $689

2b. 21 hot dogs

3a. $1,350

3b. $6,450,000

1.8

1a. $972

1b. 640 viruses

2a. $7,000,000

2b. 900,000 pounds

3a. 1,573 bottles

3b. 126 miles

1.9

1a. 30,820

1b. 6.9 ohms

2a. 1.56 ohms

2b. 4.60 ohms

3a. 250 square centimeters

3b. 30 centimeters

2.1

1a. $80.64

1b. $14.40

2a. $3.96 per pound

2b. $2.60 per pound

3a. 10 gallons at $2.48 per gallon, 5 gallons at $2.24 per gallon

3b. 12 pounds of cashews

2.2

1a.	eighty tickets	1b.	70 adult tickets, 80 children's tickets
2a.	90 hardcover books, 110 softcover books	2b.	65 plaid ties, 84 solid ties

2.3

1a.	100 gallons of California wine, 100 gallons of French wine	1b.	8 ounces of 16% gold, 24 ounces of 28% gold
2a.	125 gallons of pure peroxide	2b.	6 quarts of pure water
3a.	5 quarts	3b.	8 liters

2.4

1a.	$2,000 at 6%, $7,000 at 7%	1b.	$5,000 in stocks, $6,000 in bonds
2a.	5% U.S. Treasury Bonds, 9% state bonds	2b.	Corporation A pays 5%, Corporation B pays 6%

2.5

1a.	240 cents	1b.	450 cents
1c.	$(2x + 3)10$ cents or $20x + 30$ cents		
2a.	32 dimes, 54 nickels	2b.	120 nickels, 235 dimes
3a.	6 dimes, 5 quarters	3b.	6 nickels, 12 dimes, 13 quarters
4a.	eight $0.20 stamps, four $0.30 stamps, nine $0.34 stamps	4b.	12 nickels, 8 dimes, 6 quarters

Chapter 3

3.1

1a.	$1,104.33	1b.	$12.78
2a.	6.9	2b.	12
3a.	5, 9, 11, 12, 15	3b.	7, 11, 14, 15, 19, 21
4a.	4.8	4b.	3.375
5a.	3	5b.	6
6a.	30	6b.	3

3.2

1a.	57 ft 8 in.	1b.	128 ft 4 in.
2a.	1 ft 5 in.	2b.	28 ft 6 in.
3a.	Width = 8 ft, Length = 12 ft	3b.	Width = 5, Length = 20

3.3

1a.1.	3,740 ft	1a.2.	8.5 minutes
1b.1.	2,268 ft	1b.2.	225 seconds
2a.1.	56 ft	2a.2.	$16.24
2b.1.	304 ft	2b.2.	$407.36

3a.1.	24 ft	3a.2.	$3.68/ft
3b.1.	123 ft	3b.2.	$2.74/ft
4a.	2	4b.	14

3.4

1a.	$427.14	1b.	$2.74
2a.	10.25	2b.	16.35
3a.	19	3b.	19.6
4a.	base = 8, each equal side = 13	4b.	base = 11, each equal side = 14
5a.	base = 15, each equal side = 13	5b.	base = 17, each equal side = 19
6a.	14.8	6b.	25.7
7a.	14.52	7b.	55.2
8a.	6	8b.	7

3.5

1a.	56.8	1b.	40.6
2a.	$\overline{KM}$ = 13.1	2b.	$\overline{CD}$ = 13
3a.	$\overline{EH}$ = 5	3b.	$\overline{AD}$ = 13
4a.	105.6	4b.	73.2
5a.	18.7	5b.	$\overline{MN}$ = 7.4
6a.	3	6b.	4
7a.	$\overline{AB}$ = 13	7b.	$\overline{GH}$ = 22
8a.	$\overline{RU}$ = 9	8b.	$\overline{LM}$ = 10

3.6

1a.	107 ft	1b.	44 ft
2a.	31 in.	2b.	40 ft
3a.	5	3b.	4.8
4a.	9.42 ft	4b.	2,043 rotations

Chapter 4

4.1

1a.	$105.60	1b.	$375.36
2a.	335 sq ft	2b.	900 sq ft
3a.	86 ft	3b.	68 in.
4a.	$243.10	4b.	$1,128.96
5a.	$79.38	5b.	$884.51
6a.	14	6b.	20

4.2

1a.	$38.48	1b.	$37,656.32
2a.	b = 25.7	2b.	h = 9.6
3a.	18	3b.	9

4.3

1a.	$14\frac{11}{25}$	1b.	54.76
2a.	$4,704.48	2b.	$174.44
3a.	28.7	3b.	2.8
4a.	12	4b.	19

4.4

1a.	$551.61	1b.	$9,841.86
2a.	14	2b.	128
3a.	13	3b.	21
4a.	The triangle is larger by 9.	4b.	The square is larger by 88.

5a.	437 sq in.; $148.58	5b.	76 sq in.; $199.88

4.5

1a.	8,490.56 sq ft	1b.	28.26 sq ft
2a.	475 sq ft	2b.	59.4
3a.	27	3b.	76 ft
4a.	21.5 sq ft	4b.	9.12 sq ft

4.6

1a.	140	1b.	338
2a.	21	2b.	6
3a.	5	3b.	3

Chapter 5

5.1

1a.	5,070 cu ft	1b.	113.632 cubic meters
2a.	11.6 in.	2b.	16 ft
3a.	72 minutes or 1 hour 12 minutes	3b.	50 minutes
4a.	207.36 ounces or 12.96 pounds	4b.	1,555.2 ounces or 97.2 pounds
5a.	0.2 in. thick	5b.	54 in. high
6a.	5	6b.	3

5.2

1a.	39,304 cubic centimeters	1b.	17,576 cu in.
2a.	76.8 kilograms	2b.	10.1 kilograms
3a.	0.8 meters	3b.	7 ft

5.3

1a.	21 cu in.	1b.	152.1 cu in.
2a.	Pico's bar is larger by 0.29 cu in.	2b.	The prism is larger by 69.3 cu ft.
3a.	19 meters	3b.	11 ft

5.4

1a.	180.864 cubic meters	1b.	4,082 cu ft
2a.	4 ft	2b.	10 ft
3a.	The rectangular solid is larger by 12 cu ft.	3b.	180 prisms

5.5

1a.	132 cu ft	1b.	308 cu in.
2a.	9 ft	2b.	21 ft
3a.	126 seconds	3b.	471 seconds

5.6

1a.	113.0 cubic meters	1b.	268 cu in.
2a.	$r = 2$	2b.	$r = 5$ in.
3a.	57 seconds	3b.	84 seconds

Chapter 6

6.1

1a.	22	1b.	31
2a.	20, 80	2b.	29, 41
3a.	7, 21	3b.	15, 22
4a.	12, 17	4b.	12, 33

6.2

1a.	6x	1b.	6t + 2
2a.	Malka is 18; Heinrich is 30.	2b.	Guillermo is 28.
3a.	Mike is 20.	3b.	David is 8.
4a.	Mallory is 23.	4b.	Marcus is 35.

6.3

1a.	54, 55	1b.	111, 112
2a.	74, 75, 76	2b.	23, 24, 25, 26
3a.	15, 16, 17	3b.	12, 13, 14

6.4

1a.	$P + 2, P + 4, P + 6$	1b.	$7t + 2, 7t + 4$
2a.	$7x + 7, 7x + 9$	2b.	$3x^2, 3x^2 + 2, 3x^2 + 4$
3a.	42, 44, 46	3b.	78, 80, 82, 84
4a.	88, 90, 92	4b.	34, 36, 38
5a.	44 coins	5b.	16

6.5

1a.	$6R + 3$	1b.	$4t^3 + 5$
2a.	$2x + 7$	2b.	$7t^2 + 6$
3a.	37, 39, 41	3b.	69, 71, 73, 75
4a.	Yes. 83, 85, 87	4b.	91, 93, 95, 97

Chapter 7

7.1

1a.	28	1b.	30
2a.	55	2b.	99
3a.	60	3b.	78
4a.	460	4b.	648
5a.	30	5b.	1,032

7.2

1a.	869.1 miles	1b.	98.4° Fahrenheit
2a.	Minimum is 6.1 miles.	2b.	Minimum is 54.
3a.	71.3° Fahrenheit	3b.	136.8 pounds

7.3

1a.	19	1b.	7.5
2a.	75.5	2b.	568

7.4

1a.	19	1b.	85
2/3a.	no mode	2/3b.	1, 3, and 5

7.5

1a.	3,514	1b.	356
2a.	25.0	2b.	19.0

7.6

1a.	10 days	1b.	11.2 days
2a.	83.75%	2b.	97.25%
3a.	97.5 percentile	3b.	16th percentile
4a.	3,750	4b.	368
5a.	2.5 percentile	5b.	99.75 percentile
6a.	215	6b.	2,880

8.1

1a. 9, 3

1b. 22, 7

2a. smaller number = 4, larger number = 7

2b. smaller number = 3, larger number = 7

3a. 15, 45

3b. 7, 35

4a. smaller number = 3, larger number = 12

4b. smaller number = 6, larger number = 7

8.2

1a. 1 apple costs $0.36, 1 peach costs $0.52

1b. 1 banana costs $0.22 cantaloupe costs $1.30

2a. 4,000 type A batteries, 2,000 type B batteries

2b. 3,200 Perfect Printers, 1,300 Perfect Printers +

8.3

1a. 20 mph

1b. windspeed = 20 mph, time = 6 hours

1c. rate of current = 3 mph, time = 12 hours

1d. time = 3 hours, rate of current = 2 mph

2a. windspeed = 15 mph, rate of plane in still air = 670 mph

2b. current = 2.5 mph, rate of boat in still water = 12.5 mph

2c. rate of plane in still air = 550 mph, windspeed = 50 mph

2d. rate of boat in still water = 10 mph

8.4

1a. 82

1b. 76

2a. 43

2b. 72

3a. 28

3b. 93

8.5

1a. $6,000 in real estate, $12,000 in the bank

1b. $4,000 at 7%, $7,000 at 6%

1c. $4,000 at 6.5%, $12,000 at 7%

1d. $8,000 at 7%, $19,000 at 8%

2a. $170,000 at 8%, $80,000 at 9%

2b. $150,000 at 8%, $350,000 at 7.5%

2c. $11,000 at 9%, $7,000 at 7.25%

2d. $6,000 in a bank, $9,000 in private business

9.1

1a. $9y^3/7$

1b. $3a^4/2$

2a. 7:6

2b. 3:2 or 3/2

3a. 5:4 or 5/4

3b. 20:27 or 20/27

4a. 9:5 or 9/5

4b. 2:3 or 2/3

5a. 5:4 or 5/4

5b. 4:5 or 4/5

6a. 2:1 or 2/1

6b. 3:1 or 3/1

7a. 27:34 or 27/34

7b. 1:14 or 1/14

8a. $72,000

8b. $2,250,000,000

9a.	85	9b.	1,520
10a.	5	10b.	75

9.2

1a.	27	1b.	63
2a.	$8.19	2b.	$12.48
3a.	75 cars	3b.	72 books
4a.	12	4b.	14

9.3

1a.	$\frac{4}{3}$ pounds	1b.	$\frac{1}{3}$
2a.	6.4 in.	2b.	15 ounces
3a.	9%	3b.	16%

9.4

1a.	22 lb/sq in.	1b.	8 cu ft
2a.	$31.25	2b.	7 people
3a.	9 in.	3b.	7 ounces
4a.	$5	4b.	9 hours

9.5

1a.	$2.40	1b.	6,720 pounds
2a.	0.3	2b.	0.9

9.6

1a.	6 ohms	1b.	3 ohms
2a.	0.6 square centimeters	2b.	24 centimeters

Chapter 10

10.1

1a.	$8 + 7x = 22$ or $7x + 8 = 22$	1b.	$6x - 4 = 67$
1c.	$9x - 11 = 56$	1d.	$83 - 5x = 23$
2a.	$a = 7, b = 3, c = -8$	2b.	$a = -9, b = -1, c = -9$
2c.	$a = 4, b = 0, c = 0$	2d.	$a = 5, b = 0, c = 6$
2e.	$a = 2, b = -8, c = 5$ or $a = -2, b = 8, c = -5$	2f.	$a = -4, b = -2, c = 13$ or $a = 4, b = 2, c = -13$
2g.	$a = 6, b = 3, c = -10$ or $a = -6, b = -3, c = 10$	2h.	$a = -4, b = -4, c = 5$ or $a = 4, b = 4, c = -5$
2i.	$a = 4, b = -12, c = -9$	2j.	$a = -3, b = -21, c = -6$

10.2

1a.	$x = 1, 3$	1b.	$x = -4, 2$

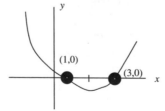

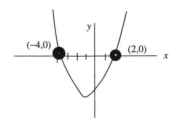

1c. $x = -2, 3$ 1d. $x = -1, 4$

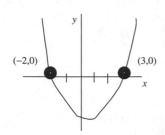 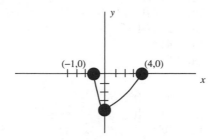

2a. $x = -3, -2$ 2b. $y = 3, -2$
2c. $z = 5, -5$ 2d. $r = -\frac{2}{3}, 4$
3a. $x = 5, -2$ 3b. $x = 3, -6$
3c. $x = \frac{1}{3}, -\frac{3}{4}$ 3d. $x = 3, -\frac{3}{5}$

10.3
1a. $28/hr   1b. 6 hr
1c. 8 cars per day   1d. $60

10.4
1a. 2   1b. 7
1c. 8   1d. 3
2a. 11, 12   2b. 9, 11
2c. 8, 10   2d. 7, 8

10.5
1a. $b = 6'', h = 11''$ 1b. $b = 14''$
2a. $l = 10''$ 2b. $w = 8''$
3a. $w = 5'', \ell = 7''$ 3b. $w = 9'', \ell = 12''$

10.6
1a. Harold takes 30 minutes alone; John takes 45 minutes alone.   1b. Vanessa takes 6 hours alone; Hilda takes 12 hours alone.

1c. Carmen takes 8 hours alone; Leshawn takes 24 hours alone.

10.7
1a. 13   1b. 8, 15, 17
1c. 6, 8, 10   1d. 26

10.8
1a. 12 ohms, 24 ohms   1b. 4 ohms, 12 ohms
1c. 2 ohms   1d. 6 ohms

11.1
1a. 15 feet   1b. 34
2a. 3 miles   2b. 150 feet
3a. 5', 12', 13'   3b. 6

11.2
1a. 4.85 feet   1b. 68.6 feet
2a. 11.05 feet   2b. 15.6 feet
3a. 53°   3b. 66°

Chapter 11

11.3

1a.	532 ft	1b.	945 ft
2a.	4.7 ft	2b.	353 ft
3a.	22,149 ft	3b.	40.1 ft
4a.	47°	4b.	33°
5a.	31°	5b.	23°

11.4

1a.	17.4 ft	1b.	3,149 ft
2a.	10.3 ft	2b.	69.6 ft
3a.	35°	3b.	69°

Chapter 12

12.1

1a.	22.5	1b.	108.8 ft
2a.	28	2b.	32 ft
3a.	Not similar	3b.	Yes. 94°
4a.	8	4b.	10

12.2

1a.	484	1b.	11,437 sq ft
2a.	35°	2b.	47°

12.3

1a.	19.4	1b.	189 miles
2a.	39°	2b.	57°

12.4

1a.	745.7 ft	1b.	319
2a.	86 miles	2b.	50.4 ft
3a.	97 ft	3b.	2.4 miles
4a.	59°	4b.	17°

Chapter 13

13.1

1a.	98	1b.	7.2
2a.	42	2b.	48
3a.	13	3b.	9
4a.	6	4b.	$\frac{1}{2}$
5a.	$2\sqrt{3}$	5b.	$2\sqrt{7}$

13.2

1a. $f'(x) = 0$
1b. $f'(x) = 0$

2a. $f'(x) = 45$
2b. $f'(x) = 23$

3a. $f'(x) = 24x^3$
3b. $f'(x) = 20x^4$

4a. $f'(x) = 12x^3 - 6x^2 + 2x$
4b. $y' = -10x^4 + 32x^3 - 9x^2 + 14x - 5$

5a. $y' = 32x^3 + 24x^2 - 12x$
5b. $f'(x) = 120x^5 + 140x^3 - 24x^2 + 72x - 14$

6a. $y' = 96x^5 + 200x^4 + 100x^3 - 72x^2 - 60x$
6b. $f'(x) = 32x^7 - 84x^6 + 54x^5 + 32x^3 - 36x^2$

7a. $f'(x) = \dfrac{2}{\sqrt{4x+7}}$ or $2(4x+7)^{-1/2}$
7b. $y' = \dfrac{5}{2\sqrt{5x-4}}$ or $\dfrac{5}{2}(5x-4)^{-1/2}$

8a. 1.2 cu in./sec
8b. 25.1 cu in./sec

9a. $70 per guest
9b. $19,990

10a. 1.8 pounds per month

10b. 43 milligrams per hour

13.3

1a. 17

1b. 1

2a. slope is −2 at $x = -4$, slope is +2 at $x = -2$

2b. slope is 9 at $x = 4$, slope is −9 at $x = -5$

13.4

1a. 0

1b. 0

2a. 36

2b. $84x^2$

3a. $72x - 4$

3b. $84x^2 - 12x + 14$

4a. $-\frac{4}{25}x^{-8/5}$

4b. $\frac{1}{3}x^{-7/3}$

5a. $300x^2 + 140$

5b. $1{,}470x^4 + 168x$

6a. 34 pounds per day

6b. 28 words per day

13.5

1a. 7.5, 7.5

1b. 15, 5

2a. 125 ft by 125 ft

2b. 225 ft by 450 ft

3a. $2\frac{2}{3}$ in. by $2\frac{2}{3}$ in.

3b. $4\frac{2}{3}$ in.

4a. 12 bikes

4b. 2,500 television sets

5a. $16

5b. $18

13.6

1a. $v(2) = 2.8$ mph

1b. $v(3) = 117$ mph

2a. $t = 3$ seconds; $v(3) = 96$ ft/sec

2b. $t = 20$ seconds; $v(20) = 640$ ft/sec

3a. $s(2) = 100$ ft; $v(2) = 18$ ft/sec

3b. $s(5) = 9{,}600$ ft; $v(5) = 1{,}840$ ft/sec

4a. $s(6) = 1{,}296$ ft; $v(6) = 312$ ft/sec

4b. $s(2) = 250$ ft; $v(2) = 157$ ft/sec

5a. $t = 6$ sec; $s(6) = 576$ ft

5b. $t = 3$ sec; $s(3) = 144$ ft

13.7

1a. $v(4) = 65$ ft/sec; $a(4) = 7.5$ ft/sec^2

1b. $v(3) = 135$ ft/sec; $a(5) = 510$ ft/sec^2

2a. $t = 9$ sec

2b. $t = 7$ seconds

14.1

1a. $x^4/4 + k$

1b. $x^8/8 + k$

2a. $x^5 + k$

2b. $(7/4)x^4 + k$

3a. $(4/5)x^5 + (1/2)x^4 - (1/3)x^3 + 3x^2 - 2x + k$

3b. $(5/4)x^4 - (4/3)x^3 + (9/2)x^2 - 2x + k$

4a. $(2/3)x^3 - (5/2)x^2 - 12x + k$

4b. $(x^4/4) - (2/3)x^3 - (5/2)x^2 + 10x + k$

5a. $(3/4)x^{4/3} + k$

5b. $(5/6)x^{6/5} + k$

6a. $8\frac{2}{3}$

6b. 60

7a. 625

7b. 315

8a. $10\frac{7}{12}$

8b. $116\frac{2}{3}$

14.2

1a. $18\frac{2}{3}$

1b. $14\frac{5}{6}$

2a. $1\frac{1}{3}$

2b. $4\frac{1}{2}$

3a. 121.5

3b. $\frac{1}{3}$

4a. 33

4b. $5\frac{1}{3}$

5a. $3{,}733\frac{1}{3}$ millions

5b. $32.6 million

| 6a. | $296,000 | 6b. | $1,204,000 or $1,204 thousands |

14.3

1a.	$-0.05x^2 + 2x + 30$	1b.	$-(0.10/3)x^3 + 40x + 25$
2a.	$101.25	2b.	$110
3a.	$1,419	3b.	$4,000
4a.	$440	4b.	$919.50
5a.	$1,600	5b.	$1,122,000

14.4

1a.	$\frac{2187}{7}\pi$	1b.	179
2a.	4.5π	2b.	$\frac{2}{3}\pi$
3a.	$17\frac{1}{15}\pi$	3b.	$4\frac{4}{5}\pi\sqrt{3}$

Solutions to Tests

Chapter 1

1.	$3,150	2.	$384.75
3.	$c = 3A - a - b$	4.	$h = \frac{V}{\pi r^2}$
5.	$C = 25°$ Celsius	6.	$F = 230°$ Fahrenheit
7.	$1,680	8.	$10,947
9.	6%	10.	364 miles
11.	4,226.75 miles	12.	2:30 P.M.
13.	82 minutes	14.	$115
15.	31,800 meters	16.	1.458 grams
17.	1,170,000 cases	18.	$12
19.	1.18 ohms	20.	50 in.

Chapter 2

1.	$178.88	2.	$3.84 per pound
3.	4 pounds at $3 per pound	4.	70 members, 110 nonmembers
5.	14 hard drives, 22 computers	6.	169 liters at 4%, 91 liters at 8%
7.	160 ounces of water	8.	$2,500 in Sterling, $6,500 in HGT
9.	9 nickels, 13 dimes, 27 quarters	10.	15 quarts
11.	5 dimes, 7 nickels, 6 quarters	12.	eight 10¢ stamps, sixteen 20¢ stamps, seventeen 34¢ stamps
13.	125 gallons of pure peroxide	14.	$485
15.	30 type A nails, 25 type B nails	16.	12 pounds of mints, 18 pounds of chocolates
17.	49 calculators at $18.50 each, 54 calculators at $15.75 each	18.	36 quarts
19.	real estate pays 9%, mortgage pays 7%	20.	300 cents

Chapter 3

| 1. | $1,005.43 | 2. | 6, 8, 11, 14, 17 |
| 3. | 4 | 4. | 3 |

5. 236 ft 2 in.
6. Width = 15 ft,
   Length = 25 ft
7. 33
8. 54.7
9. 23.79
10. 6
11. 5
12. 4
13. 289 ft
14. 46 ft
15. 24
16. 2,400 revolutions
17. base = 31, each equal side = 18
18. $\overline{MN}$ = 13

1. $198.36
2. 373 sq ft
3. $1,764
4. $69.12
5. 8
6. $770.64
7. 97
8. 29
9. $29,378.16
10. 16
11. $1,397.76
12. 36
13. 7
14. The triangle is larger by 41 square units.
15. 153.86 sq ft
16. 8
17. 21.5 sq ft
18. 130.5
19. 39
20. 8

1. 3,855.6 cu in.
2. 8.3 in.
3. 31.5 minutes
4. 2,520 ounces or 157.5 lbs
5. 2 in.
6. 4,096 cu cm.
7. 64.8 kilograms
8. 0.9 meters
9. 230.1 cu ft
10. Felix candy bars are 0.13 cu in. larger.
11. 20 meters
12. 4 meters
13. 352 cu in.
14. 3 ft
15. 14 seconds
16. 9 meters
17. 27 cones
18. 268 cubic meters
19. 4
20. 56 seconds

1. 12
2. 15, 30
3. 6, 30
4. 16, 36
5. $7g + 4$
6. 25
7. 19
8. 36
9. 33, 34, 35, 36
10. 7, 8, 9
11. 82, 84, 86
12. 14, 16, 18
13. 34
14. 17, 19, 21, 23
15. 37, 39, 41
16. $4x + y + 2, 4x + y + 4$
17. $4p - 1, 4p + 1$

1. 45
2. 230
3. 30
4. 588
5. 26
6. 5.1
7. 38 mpg
8. 13.6
9. 5
10. 120.5
11. 5
12. 1, 4, 7, 9
13. 4,046
14. 3.5
15. 3.8
16. 81.5%

17. 84th percentile   18. 750
19. 0.25 percentile   20. 200

Chapter 8

1. 9, 23
2. 5, 8
3. smaller number = 7, larger number = 11
4. smaller number = 2, larger number = 8
5. one shirt costs $23, one tie costs $5
6. one blouse costs $18, one skirt costs $38
7. 320 softcover books, 80 hardcover books
8. 210 wire-frame eyeglasses, 40 plastic-frame eyeglasses
9. 6 mph
10. 650 mph
11. rate of the current is 3 mph, rate of the boat is 27 mph
12. 3 mph
13. 93
14. 67
15. 47
16. 64
17. $5,000 at 7%, $9,000 at 6%
18. $15,000 in real estate, $35,000 in U.S. Treasury Bills
19. $7,000 in a mutual fund, $3,000 in common stocks
20. $9,000 in Turkish bonds, $41,000 in British bonds

Chapter 9

1. $9a^2/2$
2. 15/4
3. 8/1
4. 13/15
5. 45 ounces
6. 75 freshmen
7. 11
8. $12.24
9. 16
10. 1.5 or $1\frac{1}{2}$ ounces
11. 0.3
12. 8%
13. 34 lb/sq in.
14. 6,200
15. $30
16. 3 in.
17. 2.5
18. 24,000 apples
19. 3.125 ohms
20. 5 square centimeters

Chapter 10

1.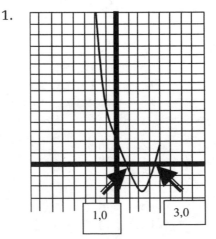

$x = 1, 3$

1,0   3,0

2. $x = -\frac{1}{2}, 3$

3. $x = 1.9, -3.4$

4. 90 beanbags per day
6. 8
8. 17, 19, 21
10. 6
12. Base = 18 inches,
    Height = 12 inches
14. Base = 12, Height = 19

16. r_1 = 12 ohms,
    r_2 = 6 ohms
18. Carlos takes three
    hours to complete the
    job alone.
20. AB = 15

5. 1,800 radios
7. 14, 15
9. 12, 14, 16
11. 21, 22, 23
13. Length = 17 in.,
    Width = 12 in.
15. Width = 5,
    Length = 8
17. R = 2 ohms,
    r_1 = 6 ohms
19. The ladder is 15
    feet long.

Chapter 11

1. 34 ft
3. 116.1 ft
5. 9,964 ft
7. 6
9. 162 ft
11. 10.6 ft
13. 43°

2. 5.2 ft
4. 9 ft
6. 67°
8. 40°
10. 448 ft
12. 2,347 ft
14. 24

Chapter 12

1. 5 ft
3. 72
5. 8
7. 1,174 ft
9. 361 ft
11. 458 million miles

2. 382 ft
4. 78°
6. 2,097 sq in.
8. 22°
10. 44°
12. 38°

Chapter 13

1. 14
3. 32.4 cu in./sec
5. 2 milligrams/hr
7. slope = −3 at x = 3/2,
   slope = 3 at x = 3
9. $120x^4 - 96x$
11. 200 ft by 400 ft
13. 38 jeans
15. 329 mph
17. 1,032 ft
19. $v(8)$ = 256 ft/sec

2. $30x^4 - 12x^2 + 4x - 9$
4. $670
6. 42
8. $84x^2 + 18x$
10. 54 people/yr
12. 3 in. by 3 in.
14. $50
16. 100 ft
18. 400 ft
20a. 832 ft/sec
20b. 352 ft/sec²

Chapter 14

1. $\frac{1}{5}x^5 + k$
3. $x^4 + \frac{2}{3}x^3 - \frac{3}{2}x^2 + 4x + k$
5. $\frac{6}{7}x^{7/6} + k$
7. 124
9. $10\frac{2}{3}$
11. 36
13. $24\frac{2}{3}$

2. $x^3 + k$
4. $\frac{2}{3}x^3 + \frac{9}{2}x^2 - 5x + k$
6. $18\frac{2}{3}$
8. $43\frac{1}{5}$
10. $30\frac{2}{3}$
12. 68
14. $558.45 thousands
    or $558,450

15. $-0.1x^2 + 6x + 200$
17. $8,880
19. $185,000

16. $2,187\pi$
18. $916
20. 2.25π

Table of Values of Trigonometric Functions

Angle	Sine	Cosine	Tangent	Angle	Sine	Cosine	Tangent
1°	0.0175	0.9998	0.0175	46°	0.7193	0.6947	1.0355
2°	0.0349	0.9994	0.0349	47°	0.7314	0.6820	1.0724
3°	0.0523	0.9986	0.0524	48°	0.7431	0.6691	1.1106
4°	0.0698	0.9976	0.0699	49°	0.7547	0.6561	1.1504
5°	0.0872	0.9962	0.0875	50°	0.7660	0.6428	1.1918
6°	0.1045	0.9945	0.1051	51°	0.7771	0.6293	1.2349
7°	0.1219	0.9925	0.1228	52°	0.7880	0.6157	1.2799
8°	0.1392	0.9903	0.1405	53°	0.7986	0.6018	1.3270
9°	0.1564	0.9877	0.1584	54°	0.8090	0.5878	1.3764
10°	0.1736	0.9848	0.1763	55°	0.8192	0.5736	1.4281
11°	0.1908	0.9816	0.1944	56°	0.8290	0.5592	1.4826
12°	0.2079	0.9781	0.2126	57°	0.8387	0.5446	1.5399
13°	0.2250	0.9744	0.2309	58°	0.8480	0.5299	1.6003
14°	0.2419	0.9703	0.2493	59°	0.8572	0.5150	1.6643
15°	0.2588	0.9659	0.2679	60°	0.8660	0.5000	1.7321
16°	0.2756	0.9613	0.2867	61°	0.8746	0.4848	1.8040
17°	0.2924	0.9563	0.3057	62°	0.8829	0.4695	1.8807
18°	0.3090	0.9511	0.3249	63°	0.8910	0.4540	1.9626
19°	0.3256	0.9455	0.3443	64°	0.8988	0.4384	2.0503
20°	0.3420	0.9397	0.3640	65°	0.9063	0.4226	2.1445
21°	0.3584	0.9336	0.3839	66°	0.9135	0.4067	2.2460
22°	0.3746	0.9272	0.4040	67°	0.9205	0.3907	2.3559
23°	0.3907	0.9205	0.4245	68°	0.9272	0.3746	2.4751
24°	0.4067	0.9135	0.4452	69°	0.9336	0.3584	2.6051
25°	0.4226	0.9063	0.4663	70°	0.9397	0.3420	2.7475
26°	0.4384	0.8988	0.4877	71°	0.9455	0.3256	2.9042
27°	0.4540	0.8910	0.5095	72°	0.9511	0.3090	3.0777
28°	0.4695	0.8829	0.5317	73°	0.9563	0.2924	3.2709
29°	0.4848	0.8746	0.5543	74°	0.9613	0.2756	3.4874
30°	0.5000	0.8660	0.5774	75°	0.9659	0.2588	3.7321
31°	0.5150	0.8572	0.6009	76°	0.9703	0.2419	4.0108
32°	0.5299	0.8480	0.6249	77°	0.9744	0.2250	4.3315
33°	0.5446	0.8387	0.6494	78°	0.9781	0.2079	4.7046
34°	0.5592	0.8290	0.6745	79°	0.9816	0.1908	5.1446
35°	0.5736	0.8192	0.7002	80°	0.9848	0.1736	5.6713
36°	0.5878	0.8090	0.7265	81°	0.9877	0.1564	6.3138
37°	0.6018	0.7986	0.7536	82°	0.9903	0.1392	7.1154
38°	0.6157	0.7880	0.7813	83°	0.9925	0.1219	8.1443
39°	0.6293	0.7771	0.8098	84°	0.9945	0.1045	9.5144
40°	0.6428	0.7660	0.8391	85°	0.9962	0.0872	11.4301
41°	0.6561	0.7547	0.8693	86°	0.9976	0.0698	14.3007
42°	0.6691	0.7431	0.9004	87°	0.9986	0.0523	19.0811
43°	0.6820	0.7314	0.9325	88°	0.9994	0.0349	28.6363
44°	0.6947	0.7193	0.9657	89°	0.9998	0.0175	57.2900
45°	0.7071	0.7071	1.0000	90°	1.0000	0.0000	

Glossary

Acceleration The derivative of the velocity or the second derivative of the distance function, s, with respect to time, t:

$$a(t) = \frac{dv}{dt} = \frac{d^2s}{dt^2}$$

Amount of money remaining in an account, A (after a number of simple interest additions) $A = P + I = P + PRT$, where I = interest, P = principal, R = annual rate, T = time (in years).

Areas between curves If the area is bounded by $f(x)$ and $g(x)$ and the vertical lines $x = a$ and $x = b$ and both functions are continuous on the interval $a \le x \le b$ and $f(x) \ge g(x)$, then the area between f and $g = \int_a^b f(x)dx - \int_a^b g(x)dx = \int_a^b \{f(x)dx - g(x)\}dx$.

Arithmetic mean Indicated by $\overline{x}$, read "x bar."

$$\overline{x} = \frac{\displaystyle\sum_{i=1}^{n} x_i}{n} = \frac{x_1 + x_2 + x_3 + \cdots + x_n}{n}$$

where n = the number of pieces of data.

Arithmetic Progression To find any term, a_n, in an A.P.: $a_n = a_1 + (n-1)d$, where a_1 = the first term, n = the order of the term, d = the common difference. To find the sum, S, of an A.P.: $s = \frac{n}{2}(a + l)$, where a = the first term, l = the last term, n = the number of terms.

Celsius/Fahrenheit formula $C = (5/9)(F - 32)$, where C = Celsius, F = Fahrenheit.

Circle
Area, A: $A = \pi r^2$, where π = 3.14, r = the radius
Circumference, C: $C = \pi D$, where π = 3.14 and D = the diameter

Cone
Volume, V: $V = (1/3)\pi r^2 h$, where π = 3.14, r = the radius of the circular base, h = the height of the cone

Cosine In a right triangle, if we compare the side adjacent to an acute angle with the hypotenuse, we have a ratio called the cosine (*cos* in abbreviated form):

$$\cos \angle = \frac{\text{Adjacent side}}{\text{Hypotenuse}}$$

Cube
Volume, V: $V = e^3$, where e = one edge

Cylinder
Volume, V: $V = \pi r^2 h$, where π = 3.14, r = the radius of the circular base, h = the height of the cylinder

Derivatives The derivative of a function is a second function that depicts the rate of change of the dependent variable in relation to the rate of change of the independent variable.

$$\frac{d}{dx}(a) = 0, \text{ where } a \text{ is a constant}$$

$$\frac{d}{dx}(yz) = y\frac{d}{dx}(z) + z\frac{d}{dx}(y) \text{ [product rule]}$$

$$\frac{d}{dx}(x) = 1 \qquad \frac{d}{dx}(x^m) = mx^{m-1} \text{ [power rule]}$$

$$\frac{d}{dx}(ay) = a\frac{d}{dx}(y)$$

$$\frac{d}{dx}(y^m) = my^{m-1}\frac{d(y)}{dx} \text{ [generalized power rule]}$$

Distance formula $D = rt$, where D = distance, r = rate, t = time

Fundamental Theorem of Calculus If f is continuous on $[a, b]$ and $F(x) = \int f(x)dx$, then $\int_a^b f(x) = F(b) - F(a)$. To find the area under the curve, we must first integrate the function, substitute the upper and lower limits into the new function, and then subtract the two limits.

Geometric Progression To find any term, a_n, in a G.P.: $a_n = a_1 r^{n-1}$, where a_1 = the first term, r = the common ratio, n = the order of the term. To find the sum, S, of a G.P.:

$$S = \frac{a_1 r^n - a_1}{r - 1}$$

where a_1 = the first term, r = the common ratio, n = the number of terms in the G.P.

Integration Technically, integration is the reverse of differentiation. Graphically, integration is the area under a curve, which represents a continuous function.

Law of Cosines $a^2 = b^2 + c^2 - 2bc\cos A$, where a, b, and c are sides in any triangle and $\angle A$ is the angle between sides b and c.

Law of Sines

$$\frac{a}{\sin A} = \frac{b}{\sin B} = \frac{c}{\sin C}$$

where a, b, and c are sides in any triangle.

Limits Theorems on Limits: In the following theorems, $\lim_{x \to a} f(x) = A$, $\lim_{x \to a} g(x) = B$.

1. If $f(x) = A$ (a constant), then $\lim_{x \to a} f(x) = A$.

2. $\lim_{x \to a} c \cdot f(x) = c\lim_{x \to a} f(x) = c \cdot A$.

3. $\lim_{x \to a}\{f(x) \pm g(x)\} = \lim_{x \to a} f(x) \pm \lim_{x \to a} g(x) = A \pm B$.

4. $\lim_{x \to a}\left[\frac{f(x)}{g(x)}\right] = \frac{\lim_{x \to a} f(x)}{\lim_{x \to a} g(x)} = \frac{A}{B}$, $B \neq 0$.

Marginal cost Economists have borrowed the idea of the derivative from calculus and have applied it to economics. Thus, if $C(x)$ is the cost of producing x units, $C'(x)$ is the cost of producing that one extra unit and is called the marginal cost. The marginal cost is the instantaneous rate of change of C with respect to x, $C'(x)$.

Marginal revenue $R'(x)$, the marginal revenue, is the revenue earned by selling that one extra unit. It is the instantaneous rate of change of the revenue, $R(x)$, with respect to the number of units sold, x.

Minimum and maximum points When the second derivative of a function is positive over an interval, the first derivative is increasing and the graph of the function is concave upward—a local minimum. Analogously, when the second derivative of a function is negative over an interval, the first derivative is decreasing and the graph of the function is concave downward —a local maximum.

Measurements

English Units	Metric Units
Distance	
1 foot = 12 inches	1 meter = 100 centimeters
1 yard = 3 feet	1 meter = 1,000 millimeters
1 yard = 36 inches	1 kilometer = 1,000 meters
Mass	
1 pound = 16 ounces	1 kilogram = 1,000 grams
1 ton = 2,000 pounds	
Volume	
1 quart = 2 pints	1 liter = 1,000 milliliters
1 gallon = 4 quarts	

Median The middle number in an ordered set of data.

Mode The number that occurs most frequently in a given set of data.

Net income Gross income – Cost.

Parallelogram
   Area, A: $A = bh$, where b = the base, h = the height

Prism
   Volume, V: $V = \frac{1}{2}(b)(h_t)(h_p)$, where b = the base of the triangle, h_t = the height of the triangle, h_p = the height of the prism.

Proportion If two ratios are equal, there is a proportion. In the proportion

$$\frac{a}{b} = \frac{c}{d}$$

a, b, c, and d are, respectively, called the first, second, third, and fourth terms of the proportion. The two outer terms (a and d) are called the extremes, while the two inner terms (b and c) are called the means.

Pythagorean Theorem In a right triangle, if a represents one of the legs, b represents another leg, and c represents the hypotenuse, the following relationship holds true: $a^2 + b^2 = c^2$.

Quadratic equation A quadratic equation in one unknown is an equation in which the highest power of the unknown is the second power. In the general quadratic equation, $ax^2 + bx + c = 0$, a, b, and c are real numbers and $a \neq 0$; a and b are known as the coefficients in the equation while c is the constant. The quadratic formula gives us the two roots of the quadratic equation:

$$x = \frac{-b \pm \sqrt{-b + b^2 - 4ac}}{2a}$$

Ratio A comparison of two numbers.

Rectangle
   Perimeter, P: $P = 2l + 2w$, where l = the length, w = the width
   Area, A: $A = bh$, where b = the base, h = the height

Rectangular solids
   Volume, V: $V = lwh$, where l = the length, w = the width, h = the height

Resistances
   If there are two resistors, r_1 and r_2, in a parallel circuit, the joint resistance, R, is given by the formula

$$\frac{1}{R} = \frac{1}{r_1} + \frac{1}{r_2}$$

Sigma The Greek letter sigma, Σ, is a shorthand summation notation.

Similar triangles If triangles are similar, their corresponding angles are congruent and their corresponding sides are in proportion.

Simple interest $I = PRT$, where I = interest, P = principal, R = annual rate, T = time (in years)

Sine In a right triangle, if we compare the side opposite an acute angle with the hypotenuse, we have a ratio called the sine (*sin* in abbreviated form):

$$\sin \angle = \frac{\text{Opposite side}}{\text{Hypotenuse}}$$

Sphere

Volume, V: $V = (4/3)\pi r^3$, where $\pi = 3.14$, r = the radius

Square

Perimeter, P: $P = 4s$, where s = one side

Area, A: $A = s^2$, where s = one side

Standard deviation

$$s = \sqrt{\frac{\sum_{i=1}^{n}(x_i - \bar{x})^2}{n}}$$

where $\bar{x}$ = the median.

Tangent In a right triangle, if we compare the relationship of the side opposite an acute angle with the side adjacent to that angle, we have a ratio called the tangent (*tan* in abbreviated form):

$$\tan \angle = \frac{\text{Opposite side}}{\text{Adjacent side}}$$

Trapezoid

Area, A: $A = \frac{1}{2}h(b_1 + b_2)$, where h = the height, b_1 = one base, b_2 = the second base

Triangle

Area, A: $A = \frac{1}{2}bh$, where b = the base, h = the height

$A = \frac{1}{2}bc \sin A$, where b and c are sides of any triangle and A is the included angle

Perimeter of an isosceles triangle, P:

$P = 2s + b$, where s = one of the equal sides, b = the base

Perimeter of an equilateral triangle, P:

$P = 3s$, where s = one side

Variation

Direct: If we have two variables, x and y, and both increase or decrease in the same ratio, we say that the two variables vary directly:

$$x = ky$$

Inverse: If we have two variables, x and y, and when one increases, the other decreases, we say that the two variables vary inversely:

$$xy = k$$

Joint variation: Joint variation is direct variation involving three variables. x varies jointly as y and z:

$$x = kyz$$

Combined: Combined variation is a combination of direct (or joint) variation and inverse variation. y varies directly as x and inversely as z:

$$y = \frac{kx}{z}$$

Velocity The first derivative of the distance function, s, with respect to time, t:

$$v(t) = \frac{ds}{dt}$$

Index

NOTES

NOTES

NOTES

NOTES

NOTES

NOTES

NOTES